Theatres of Dust

Linda Hassall

Theatres of Dust

Climate Gothic Analysis in Contemporary Australian Drama and Performance Landscapes

palgrave
macmillan

Linda Hassall
Griffith University
Nathan, QLD, Australia

ISBN 978-981-16-6158-7 ISBN 978-981-16-6159-4 (eBook)
https://doi.org/10.1007/978-981-16-6159-4

This Palgrave Pivot imprint is published by the registered company Springer Nature Singapore Pte Ltd.
The registered company address is: 152 Beach Road, #21-01/04 Gateway East, Singapore 189721, Singapore

To the ones who keep me grounded, Stephen Rowan, Isabella Daisy, Indigo Rose and Thea Lily.

Preface

I have always loved the landscape. The Australian landscape. The stories of it I heard as a child growing up enthralled me. The dry and dusty Central Queensland town of my childhood was the sort of place that embedded such stories, narratives, vignettes, and anecdotes in its working-class socio-cultural milieu. I was mesmerized by the country and the tragedy of those who worked in it or battled to survive it and its 'bastard' weather. The heat, the smell, the dust, the distance of that landscape has never left me. That landscape of my childhood calls to me still and I try to capture my response to its call in my writing:

> She is always behind me … the distance. … she smells red. Her colors blow the guilt winds that whisper history. The winds reach across the distance and coat me in the dust of the landscape of *my* past … and I almost smell the thin, raggedy girl wearing hand-me-down clothes, barefoot in the dirt. Long blond hair tangled down her back, her eyes always on the horizon…waiting … That girl is still within me … she is sometimes 'woman' in my work … She fears that in the centre, green is almost forgotten … a distant memory layered in dirt. A sky the colour of kerosene … of day-old bruises, streaked in nicotine glazed cloud turns to bitter memory, as night battles for a place in the scalded desert. At the back of me is always the memory of the land breathing the distant, dry smell of the bones of those who have come before me … their names never noted in plaques, or scratched into tree stumps, or chiselled on the face of monuments …

> *sshhh* … A scalding national anthem whispers as the land cracks but doesn't bleed. In her menopause, she spreads herself open for the rain that doesn't come …

My memory reveals a sun-baked landscape of (mis)-remembrances of place, a dusty (ir)-real time wherein a vivid physical and psychological landscape tapestry was, even then, unravelling. Playwrighting and performance making now provides me with the opportunity of being able to grapple with the irresistible pull of these common thematic threads and I attempt to grasp that landscape by focussing my artistic and dramatic imagining on place, geography, distance, space, belonging, salvation, damnation, dust. I am now a long way from my childhood and that (ir)-real landscape, its nostalgic memory, is vanishing. The winds reaching across the distance are more vicious, whispering a different story as the land transforms and the climate ravaged future is becoming more and more visible through the dust.

Like what is buried deep within my own memory, in the Australian nation context, the ancient and ethereal landscape has always narrated a space wherein the country itself plays out like a theatrical production with, dust, erosion and increasing catastrophic weather events the lead protagonists … *A scalding national anthem whispers as the land cracks but doesn't bleed*. Despite being an arid country that has always suffered extensive drought, devastating cyclones and catastrophic bush fires, Australia can still be described as a country that is largely un-polluted and underpopulated. It is perhaps this paradox that lulls us into a national lethargy wherein we boast—in our self-deprecating Australian tone—*it* (climate change) is something that might happen … sometime in the future and when it does … we'll deal with it, like we've always dealt with weather in this 'mongrel' country – battle it or be beaten down by it. The Australian landscape is, and always has been, a chimerical beast. It is now a site of protest and a subject of politicized debates over voracious resource extraction and ineffective land management practices. Economic stability is at the heart of environmental discourse, even as the mining industry and the farming and export culture are identified as contributing to escalating climate change … *in the centre, green is almost forgotten … a distant memory layered in dirt*. It is ironic that while we are all aware of global warming, it may be alleged that many Australians, remain either apathetic, resigned or nonplussed about scientific response to and political rhetoric about, climate change. Indeed, it has been the extreme landscape and its corresponding weather conditions that have defined and framed the popular Australian identity to the rest of the world, since colonization. Much of the artistic and scholarly work that makes up the Australian cultural and literary canon, mine included, is of and about the country, defined by the Australian landscape—how it affects the psychology of a nation, the psychology of character and corresponding character and social behaviours.

As a playwright and contemporary performance maker, I acknowledge climate science as the basis for the questions I explore through the powerful cultural platform of theatre. I am well aware that the scientific community is the group with the most up to date information on climate change. Scientists have the unenviable role in predicting the escalation of climate change, whereby often the information communicated may be construed as conflicting or confusing for the broader Australian population even as communities reel as lives and livelihoods are being taken by the voracious climate change beast. There is no doubt that the wider scientific community has continued to communicate and debate climate change implications and our careless human impact on environmental degradation for many years. The plays discussed in this book in many ways respond to the important work of climate scientists.

Australian theatre-makers are contributing to the climate discussion from unique cultural and artistic perspectives. Playwrights, performance makers, production and technical teams are using the immediacy of the theatre to engage communities through communal, interactive, or cultural explorations about the vast quantities of opposing evidence and abstract statistics concerning climate change. While scientific data dissemination may in some cases alienate the public and even render some of them frightened or helpless in the face of such overwhelming problems, theatre can uniquely explore the scientific data from personal, emotional, and psychological perspectives. Theatrical engagement is an intimate encounter between spectator and performance or play and reader and can cut through the confusion to make the unfathomable, *relatable*—relatable to individuals, communities and the broader public. In doing so, the dramatic and theatrical platforms can use, select, organize and disseminate critical climate information to: enhance general understanding of climate change, to investigate politicized debates concerning carbon emission and even pose hypotheses about climate crisis mitigation. Significantly, it can explore these concerns through our own unique ways of storytelling, a form in which the country, like it always has, becomes a central antagonist on the nation stage.

The dramatic form framing the discussion in this book draws significantly on the Australian Gothic … *At the back of me is always the memory of the land breathing the distant, dry smell of the bones of those who have come before me.* The contemporary emergence of the Gothic style in Australian drama suggests that contemporary playwrights and theatre makers are embracing elements of European and Southern American Gothic traditions and adapting them to their immediate socio-cultural context. The Gothic form enables me and the playwrights included, to pose dramatic and theatrical inquiries though landscape, and environmental and atmospheric investigations that are becoming more and more focussed on how climate change is affecting the uniquely beautiful country that we have projected to the wider world for the last two centuries. The theatrical works discussed throughout all explore (from varying perspectives) how climate change is an abstract, intangible beast, a contemporary monster of our own making. These works narrate how it has reared its ugly head, spewing fire and flood fury while simultaneously ravishing the landscape with extreme drought and erosion. They depict the landscape itself as a Gothic paradox of reversals. The Australian landscape and our place in it are explored through these works, from what I identify in this book as an *Australian Climate Gothic* perspective. The Australian Climate Gothic claims that Australian landscape is transformed, acclimating into what I further describe as climate-scape, a physical and psychological space which claims that our enduring ability to beat the odds and battle everything the landscape throws at us is a heroic memory, a disappearing myth belonging to the popular colonial ethos of the past.

I have always loved this landscape. The Australian landscape … *She is always behind me … the distance. … she smells red. Her colours blow the guilt winds that whisper history.* It is obvious I have a strong sense of place attachment. Therefore, it is important for me to acknowledge that the discussions within concerning white inheritance, legacy and identity, are formulated from my privileged position as a white, middle-class female artist and scholar, and as such I respectfully acknowledge that many First Nations and Torres Strait Islander Peoples do not share a sense of harmony with non-indigenous Australians about the land.[1] Further, it is a simple fact that self-preservation in a politicized climate affected landscape suggests that many Australians are cautious about environmental issues taking up too much time and space, especially in the wake of COVID 19, where the population does not recognize how environmental consciousness will kick start a flailing economy reliant on mining and agricultural and meat export. The varied dramatic imagination(s) that form the content analysis in this book attempt to provide some insight into how environmental analysis can be shared, how theatre can speak to people in a way that science cannot and further, how the theatrical imagination can offer a platform to situate important climate discourse.

Nathan, Australia Linda Hassall

Note

1. See Sarah Maddison's (2011) book *Beyond White Guilt: The Real Challenge for Black-White Relations in Australia*, which discusses ways modern Australians can right historical wrong and acknowledge collective responsibilities to develop a revitalized version of the nation and national identity.

Acknowledgements

I would like to thank Griffith University for giving me the time and space to write this book. This book specifically acknowledges Australian artists, playwrights, theatre makers and theatre scholars who are responding to climate change in their own unique way. Every day, the work of these theatre makers reminds me of how this beautiful country, while hurting, remains the inspiration for a contemporary analysis of the nation space, the landscape in which we are inextricably tied and the sunburnt country we call home. I would like to specifically thank the playwrights who agreed to me analysing their work in this book: Sarah Hamilton and Justine Campbell, Kathryn Marquet, Shaun Charles, Jackie Smith, Stephen Carleton and Elaine Acworth. Their dedication to the exploration of the climate-affected nation space is significant in generating a new genre of Australian drama, identified in this book as the Climate Gothic. I would also like to acknowledge those playwrights and theatre makers whose work I haven't been able to include. Further, I acknowledge those behind-the-scenes technical production teams, eco-scenographers and companies who are doing their best to respond to climate change in the Australian contemporary performance sector. Their efforts to reduce damaging energy and resource wastage in theatre production practices are advocating for an Australian sustainable theatre practice culture. I would especially like to acknowledge Stephen Rowan's research into sustainable production technologies. His efforts in promoting energy efficient and sustainable theatre production practices have had a huge impact on my own artistic practice and in my promotion of ecological stewardship in emerging theatre makers. Finally, a big thank you to Dr Natalie Lazaroo for all her support in the proofing of the manuscript.

Contents

Abbreviations

Climate Gothic	Australian Climate Gothic Drama
Climate-Scape	Climate Affected Landscape
Myth-Scape	Mythic Landscape

Notes on Terminology

(Mis)-Remembrance, (Ir)-Reality and Subjectivity

(Mis)-Remembrance and (ir)-reality are personal terms used to excavate and define my memories of the outback landscape. My memories of the past and what it represents, in many ways defines me as woman, artist and scholar. The landscape I remember is simultaneously real and aesthetically, something other than real—enhanced by my subjective associations to it and by my experiences within it. While not being historically true or indeed accurate to geography, it is true and accurate to me. Consequently, (mis)-remembered and (ir)-real landscape explores the tension between actual geographic landscape and biographical constructs of landscape phenomena, flavoured by my personal experiences of growing up in it (see analysis of my own plays in Chapters 2 and 3).

White Inheritance

White inheritance is a term used in this study specifically to identify tensions associated with Australian land ownership/acquisition and custodial heritage of land held by white Australians of English and/or European descent.

CHAPTER 1

Theatres of Dust

Abstract This chapter introduces how Australian playwrights are exploring Australia as a climate-affected space. It offers two concepts, which underpin the discussion throughout the book. The first concept is that of *climate-scape*. Climate-scape describes the contemporary climate-affected Australian landscape and suggests that dramatists are reimagining enduring Australian landscape legacies through climate change lenses. Climate-scape narrates a space wherein Australia itself plays out like a theatrical production, with dust, erosion, increasing catastrophic weather events and politicized debates over resource extraction, land management and climate change as the lead protagonists. The second concept extends the definition of the Australian Gothic into *Australian Climate Gothic Drama*. Described as *Climate Gothic*, the term suggests that the popular landscape of the past is transforming, disappearing as the climate change fury increases extreme weather conditions. The discussion proposes that *Climate Gothic* plays utilize theatrical atmospheres, scenographic imagery and dramatic imaginings about the climate-affected planet as provocations for discourse or change. Finally, this chapter introduces the plays discussed in this book.

Keywords Climate Change · Climate-scape · Australian Climate Gothic

L. Hassall, *Theatres of Dust*,
https://doi.org/10.1007/978-981-16-6159-4_1

Climate change is becoming the popular vehicle for dramatic reflection and critique on Australian society's ideals, values, way of life, on its history of material consumption and on the tensions between individual self-fulfilment and respect towards the other-than-human. As a phenomenon, the climate gives expression to the nation's anxieties about natural disasters and provides a framework for imagining the uncertain future we might leave the generations to come. Amidst the calamity and confusion associated with escalating climate change, some Australian dramatists are utilizing their distinctive skills in deciding what the climate story is, how we might respond to it and how to tell it in the cultural realm. I refer to the works of these playwrights as contributing to the *Theatres of Dust*.

Theatres of Dust suggest that the detrimental impacts of climate change are becoming a popular theme in Australian performance and playwriting[1] culture. Contemporary performance makers and playwrights are investigating how historical relationships with coal and cattle, habitat, cultural geography, non-human species and consumer and material practices have informed new descriptions of what is an almost unrecognizable climate-affected space. To provide a national context, climate change is having a devastating effect on the Australian landscape historically narrated to the world through unique images of an ancient and beautiful, land. All that is distinctive about the country and its character have perpetuated a mythic space to the rest of the world since colonization (see Chapter 4). However, the mythic landscape is no longer out there—if it ever was—and the performances and plays discussed throughout this book bring to the fore new climate-affected landscapes, acknowledged within as *climate-scapes*. Climate-scape as a term draws on Chaudhuri's (2002) discussion[2] on landscape theatre, and further suggests that theatrical climate-scape exploration enables material and egocentric analysis about human impacts on the country of Australia. Climate scientists observe that conditions on the Australian continent are already more unpredictable than in the rest of the world, becoming more complex and more variable as the planet warms and human-induced climate change affects a country that Robin (2010) describes as already under stress. As a dramatic concept, climate-scape acknowledges scientific climate variables in the language, materiality and imagination of the theatre. It embraces both culture-scape and landscape, and identifies a discourse of analysis between land, habitat and species management and the impact of an economic ethos that has resonated throughout the national Australian ideology since colonization. The dramatic discourse reimagines enduring

Australian cultural legacies that have previously defined the Australian relationship to the beloved Bush, Beach or Outback. Climate-scape suggests that these enduring legacies are vanishing, and playwrights and performance makers are exploring the part we humans have played in this present ecological situatedness.

As the effects of climate change on the country accelerate, the Australian landscape is undergoing significant change, becoming *unheimlich*—unfamiliar, strange. While Australians are aware of climate change and what it is doing to the country and to lives and livelihoods, climate change is still a seemingly abstract concept to many, something that happens ... out there ... an intangible, esoteric beast identified through the escalation of harsh weather, mining and habitat destruction. Sea levels are rising at an alarming rate. Coastlines are washing away and beachfront homes in much sought after Australian coastal cities such as the famous Gold Coast are collapsing into the Pacific. The damage to the environment and native habitat is escalating with the ADANI mining expansion into Queensland and the resulting ghosting of the Great Barrier Reef that has recently experienced its third major bleaching event. Yet the mining industry is the beast that keeps the economy afloat even as it destroys the landscape we have prided ourselves on perpetuating to the wider world. Most Australians have experienced firsthand the devastating effects of long-term drought, destructive floods and/or catastrophic and frequent bush fire events that are destroying our once beloved land of dazzling beaches, sweeping plains and mythic distances. What were once considered as one in one-hundred-year weather events are now happening daily, weekly, monthly.

The climate-scapes informing the plays and performances explored in this book include elements of Gothicism and as such, the plays analysed are situated under the umbrella term of *Australian Climate Gothic Drama* (hereafter referred to as Climate Gothic). The works examined utilize theatrical atmospheres, scenographic imagery and dramatic imaginings to provoke questions concerning the climate-affected world, specifically in an Australian context. In Australia there is not yet a large body of dramatic fiction or performance that can be classified as dystopian, anti-utopian and apocalyptic as in contemporary Australian literary studies and it is hoped that this book opens up a space for discussion and analysis of theatrical works that can be classified as such.[3] The climate-scapes discussed, imagine and describe liminal paradoxical spaces that exist between perceptions of what is absent and what is present, between what is past and what is future, between climate and human and non-human nature. The climate-scape paradox, therefore, is a Gothic one and can be identified as the contemporary quandary that explores the horrors associated with (post)modernity through various sociological and environmental lenses. In recent Australian drama of the late 1990s and early twenty-first century, the Gothic is the space where the Nation's collective repression is brought to the forefront of contemporary political culture to expose the Nation's infamies, to explore the Nation's ghosts and to mourn the Nation's vanishing places, spaces and species. Drawing on themes of place and space, Australian Gothic has incorporated national anxieties pertaining to colonialism and (post[4])colonialism in the popular resurgence of the Gothic form. It is worth noting that traditional Australian Gothic perspectives are viewed through thematic explorations of (national) guilt associated with colonial behaviours of terra nullius, settlement, appropriation of and enclosure of land by civilized society, and with treatment of First Nations People. Climate Gothic extends on these themes to examine how civilized European society and socially conditioned cultural groups have exploited the land they misappropriated and investigates the repercussions of that from various socio-cultural perspectives. As such the collection of plays that make up the analysis in this book engage with disappearing natural landscapes, land mismanagement, species extinction and escalating natural disaster. Many engage with new ways of identifying the Australian landscape and the (white) Australian identity by reimaging many of our beloved cultural myths. The plays included, by some of Australia's leading playwrights, situate what is specifically local or national in relation to the ever-increasing global crisis and the analysis draws on scholarship from literary, humanities, eco-critical and social science fields. The Climate Gothic plays analyzed are peculiarly Australian. Some like my own *Dust*[5] and *Salvation*, which open the discussion proper in this book, reimagine Australian experiences of white estrangement in the Australian climate-scape by questioning the colonial past and other enduring legacies that will affect our environmental future. Other plays thematically narrate the quest for survival of other than human species, e.g. *They Saw a Thylacine* and *The Dead Devils of Cockle Creek*, and others explore the tension between colonial informed white nation spaces *and* debates about our responsibility to biodiversity in the contemporary climate-scape, e.g. *The Turquoise Elephant* and *The White Earth*. Others, such as *The Flood* and *Water Wars*, examine how the escalating climate crisis affects the future of Australian families in a time of climate change and exacerbated natural disaster.

All the plays included have emerged as provocations—provocations about new kinds of disruptions in the (post) colonial Australian landscape. While each chapter in this book extends the Australian Gothic trope into the Climate Gothic discussion, it is important to note that from a contemporary perspective, they emerge from existent tensions in human–landscape interactions that are the effects of past imperial acts of governing, managing, taming and domesticating the geography. In reimagining landscape dialogues from this perspective, playwrights are successfully opening up a discourse about uncomfortable historical truths, including Australians' fraught relationship to their colonial past, to each other and to the land that is supposedly shared. All the plays are examples of how contemporary playwriting, performance and theatre making is responding to politicized Australian debates concerning the impacts of continuing drought and natural disaster on the country, and explore conflicts about climate change denial. While these issues can be argued from various political and economic platforms, the theatrical discourse included here can generate more widespread discussion or awareness about the climate challenges we may face now and in the future.

Dust (Hassall, 2015) is an apocalyptic Gothic performance text that embraces principles of what Sullivan (2015) refers to as the 'dark pastoral' in that it poses a version of climate-scape that holds little hope of rejuvenation (see Chapter 2) and as a result hypothesizes the Anthropocene's destructive experiences through the almost eradication of human species. As a Climate Gothic example, the play exposes the outcome of the Nation's collective fears when facing what Byron (2012) describes as the growing global darkness called warming. Set in a wasteland of humankind's own making, *Dust* explores the days and years after the day of the cataclysmic disaster, which heralded the end of the world as we know it. Delivered from the perspective of a last surviving community eking out an existence in the apocalyptic wasteland, the play explores conflicting positions on the climate change crisis and questions the audience as to why more was not done to avert *The Final Disaster*. Posing hypothetical experiences, *Dust* characters move in and out of time, through the chaos experienced by survivors directly after the disaster, to some remote point in the future where human beings are on the very edge of extinction. *Dust* provokes dialogues about what might happen if we continue in such a careless manner and poses questions about the environmental and ecological legacy we are leaving to those who come after us.

Salvation (Hassall, 2013) explores the birthing and dying of the Australian Nation through a series of nightmarish fragments depicting monstrous imperial activity including transportation, land appropriation, the systematic abuse and enslaving of First Nations Peoples and the development and proliferation of industries that have impacted on the contemporary climate-scape (see Chapter 3). *Salvation* charters the imaginings of the country as a border territory and investigates places and spaces where all sense of time is lost and there seems to be an absence or a vanishing of recognizable geographical and psychological cultural markers. Significantly, *Salvation* navigates the movement from Australian Gothic to Climate Gothic by exposing the traumas of a colonizing history as being directly linked to the contemporary climate-scape.

They Saw a Thylacine (2017) by Justine Campbell and Sarah Hamilton set in Tasmania also operates in two separate timelines—the present and future—and uses two very different vernaculars. The play charts the end of life and the extinction of an Australian native species (see Chapter 4). The characters articulate the care that should be taken of native species in our stewardship of place, while viscerally evoking the delicate equilibrium in our world whereby recognizable systems are tending towards decay and chaos.

The Dead Devils of Cockle Creek by Kathryn Marquet (2018), also set in the deep and dislocating Tasmanian wilderness, in many ways investigates a future wherein beliefs concerning the reconciliation between human and non-human species is no longer possible (see Chapter 4). *The Dead Devils of Cockle Creek* reveals the failure of political and environmental structures that perceive climate change and global warming as distinct impersonal threats and the repercussions of such on species extinction—specifically that of the Tasmanian Devil.

The White Earth by Shaun Charles and Andrew McGahan (2009), discussed in Chapter 5, explores how material processes of thinking about land as a divisible commodity and obsessing over its value in terms of ownership and legacy, can significantly impact on the sacred relationships of the 'other' and with non-human nature. The Climate Gothic elements are closely linked with geography of place, which suggests the environment enacts a specific interactive role in the cultural milieu. The play examines the Nation as a collective psychological space where white male decision-making processes about land and land management are stripped bare, and traditional notions of white identity and country are reimagined in the border territory between past and present. The climate-scape is both geographic and psychological, a place where dreams of home and taking ownership of the land are paramount. It's a contested space wherein the construction of that home, in true Gothic mode and described by author Andrew McGahan (2009) in the introduction, is the space for 'housing the dark dreams and desires of a male white nation [wherein] the monsters, ghosts and madmen reign' (p. 5).

The Flood (2012) by Jackie Smith evokes a climate-scape in rural Tasmania. In an old farmstead going to rack and ruin, an elderly woman and her two grown daughters view the river, which runs through the front of their property, rising during a flood (see Chapter 5). The incessant rain and the rising floodwaters set the backdrop for the escalating tension simmering between the three women. The climate-scape is menacing, the weather is fickle, flooding in the East while Bushfires rage in the West. The Climate Gothic reading sits within tension between what is familiar and simultaneously unfamiliar in the geographic surrounds. There is no reprieve for the women as roads are blocked and communication with the outside world breaks down due to the unprecedented weather event. The women have no choice but to face their demons as the family secrets thought forever hidden, emerge in a river determined to spew out its mysteries.

Stephen Carleton's *The Turquoise Elephant* (2016) narrates a sophisticated urban narrative that questions how we as humans can acknowledge we are but one generation, one species among many (see Chapter 6). Climate Gothic elements can be viewed through a post-Romantic lens wherein non-human nature is a spectacle, an abstract event that provides an opportunity to witness species extinction. The circumstances further advocate for a reintroduction of natural resource extraction as a means of contesting climate change. Australians are experiencing the highest ever recorded temperatures, and rising sea levels are swamping major cities. *The Turquoise Elephant* embraces the dominant cultural literary narrative of climate fiction (cli-fi); the impending apocalypse and the dystopian milieu that will ensue by depicting a series of absurdist operatic futuristic scenarios specifically in the cultural discourse of a major world city—Sydney. The play is important as it explores the recent status of climate change science through familiar beliefs and orientation points circulating in a recognizable cultural discourse of place, environment and experience.

The play referred to in Chapter 7, *Water Wars* (Acworth, 2011), is a significant testament to the Climate Gothic and draws on an urban Gothic perspective. The climate-scape is inner city Brisbane and offers a futuristic examination on what has become a Queensland (and Australian) backyard Grimm-like fairytale. The Australian dream of green sprinklered lawns, lush-flowered gardens, well-watered vegetables beds, and sprawling jacaranda and mango trees and cool inviting swimming pools, is ruptured through the reality of double perch[6] parched backyards, dead trees and empty pools. Significantly, the play recognizes the theatre industry's response to climate change and its potential to encourage sustainable 'green' behaviours. While the play itself is a significant example of the Climate Gothic, what is unique about this work is that it was produced almost entirely sustainably. *Water Wars* boasted a distinctive production process that agitated for sustainability both creatively and practically.

In all the works the climate-scape is the recurring motif linking elements of the Contemporary Australian Gothic with the Climate Gothic genre.[7] Often evoked as symbolic terrain, the climate-scapes brings to the fore the truth(s) about our relationships with each other and with non-human nature. Further all of the plays embrace a localized Australian Gothic sensibility, wherein the traditional 'monstrous' Gothic is restaged for a new warmed world, wherein the buried trauma of environmental and ecological foundations of white Australia are resurrected as the climate change beast raging across the ancient Great Southern Land. By dramatizing contemporary socio-environmental concerns, the Climate Gothic playwrights are transforming traditional gothic sensibilities into contemporary investigations about how and why recognizable Australian spaces and places are disappearing or are being disturbed and disrupted through climate change escalation. The Theatres of Dust offer significant cultural contributions to climate change discussions. Specifically, they are important to a contemporary exploration of the escalating and changing relationships humans have with the natural world as they hypothesize the repercussions of such on the collective national psyche.

Notes

1. I make the distinction here as some plays may not have been produced and some performances of new dramatic works may not have been published.

2. Chaudhuri, U. (2002). In E. Fuchs & U. Chaudhuri (Eds.). *Land/scape/theatre*. This book is a definitive text that proposes landscape as a necessary paradigm for understanding modern theatre's increasingly spatialized aesthetic as well as its engagement with the cultural meanings of place and space.
3. Andrew Bovell's (2008) award-winning play, *When the Rain Stops Falling*, is set in 2039. The play largely explores climate change escalation across four generations of the one family. It can be suggested that this was the first twenty-first century Australian play that kicked off climate change examination in dramatic fiction in Australia. Receiving numerous awards and achieving international recognition, the work may be described as Climate Gothic. I have not included a discussion on this significant play in this book as there are numerous scholars who have written about and analyzed this work for its climate change themes, since it premiered at the Adelaide Fringe Festival in Australia in 2008.
4. Brackets are used here to acknowledge the debate in this country as to whether Australia has indeed moved past the colonial and imperial state of mind. The negative effects of colonization are still having a significant and often drastic impact on Indigenous peoples. The social and economic impact of invasion and control of Indigenous peoples has accumulated across generations and can be acknowledged in the transmission of trauma, poverty and.

   other forms of disadvantage from one generation to the next.
5. *Dust* had two productions. The first was at the Sue Benner Theatre Brisbane City in 2014. It was adapted in 2018 by invitation for a youth audience and was produced at the Yeronga Performing Arts Centre, Yeronga High School.
6. Perch refers to a term for an Australian house block of land, usually in older, more affluent suburbs referring to 800 m square.
7. See the following chapter for a more detailed discussion on 'genre' as applied to the Gothic and the Climate Gothic.

References

Acworth, E. (2011). *Water Wars* produced at La Boite Roundhouse Theatre, Directed by Shaun Charles, 3–20 August, 2011.

Byron, G. (2012). Global gothic. In D. Punter (Ed.), *A new companion to the gothic* (Part V, pp. 369–378). Blackwell Publishing Ltd.

Carleton, S. (2016). *The turquoise elephant*. Currency Press.

Campbell, J., & Hamilton, S. (2016). They saw a thylacine. *Endangered: Three plays* (pp. 5–59). Currency Press.

Charles, S., & McGahan, A. (2009). *The white earth*. Currency Press.

Chaudhuri, U. (2002). In E. Fuchs & U. Chaudhuri (Eds.), *Land/scape/theatre*. University of Michigan Press.

Hassall, L. (2013). Salvation. In Dallas Baker & Debra Beattie (Eds.), *TEXT Journal of Literary and Creative Writing, Scriptwriting as research, 19*, 1–20.

Hassall, L. (2015). *Dust* produced at the Sue Benner Theatre. Written and Directed by Linda Hassall, September, 2015. Brisbane City, Australia.

Marquet, K. (2018). *The dead devils of cockle creek*. Playlab Press.

Robin, L. (2010). Battling the land and global anxiety. *PAN: Philosophy and Activism, Nature, 7*, 3–9.

Smith, J. (2012). *The flood*. Australian Script Centre, AustralianPlays.org.

Sullivan, H. (2015). The dark pastoral: Goethe and Atwood. *Green Letters, Studies in Eco-criticism, 1*, 47–59.

CHAPTER 2

Mutating the Australian Gothic into the Climate Gothic

Abstract The chapter develops ideas presented in the previous chapter to firstly provide a more comprehensive discussion of Australian Gothic and why the form became popular in dramatic responses to long-enduring traumas associated with colonization. It then describes how the contemporary Australian Gothic mutates immediate socio-environmental concerns into the Climate Gothic. In doing so, the chapter discusses how the Climate Gothic transforms gothic sensibilities into investigations about how Australian spaces and places are being disturbed and disrupted by the escalation of horrific drought, ravaging bush fires, devastating floods and terrifying cyclone events. The performance text that supports the discussion in this chapter, *Dust* (Hassall, 2015, 2018), is clearly situated in the Climate Gothic as it posits a future Australia as a shadow nation, an over-farmed, over-mined wasteland. The play contends that politicized arguments about climate change and subsequent lack of climate action predicts an apocalyptic future of our own making. Within the play, Australia is explored as barren terrain, the victim of the truly frightening contemporary monster, climate change, and its voracious minion, global warming.

Keywords Future Terrains · Loss · Memory

L. Hassall, *Theatres of Dust*,
https://doi.org/10.1007/978-981-16-6159-4_2

All that is Australia begins not with Black Nation or white civilization, but with the land itself. In the contemporary Australian climate, self-preservation is an experience that is just one of the many debates currently being enacted in the eternal preternatural presence which is the landscape. When discussing drama, and indeed Australian Gothic drama, as a platform for exploring important climate change themes and events, the following discussion acknowledges Carter's (2010) scholarship stating "Australia was always simply a stage, where history occurred, history a theatrical performance [where the historian-playwright] coordinated facts into a coherent sequence" (p. xiv). The great continent, its distances, its colours and its associated spatial imagining, can be acknowledged as a psychological stage on which historical memory, cultural difference, social unification and racial debate are inscribed and performed. Never more so than in the present is the ancient Australian landscape transforming, her vibrant colours now distant memories layered in dust and ash, her lifeblood reflected in the detritus of displaced people and non-human species. As a Nation, we struggle not only with climate change demons, but with our inability, even unwillingness, to exorcize them. Our sense of place is disrupted, defined by one in one hundred-year events happening monthly, weekly, daily and our experience of the popular 'mythic' Australian landscape of the past is haunting us, becoming intangible, lost in collective memory and overblown legends that have whitewashed brutal colonizing experiences. A narrative representation of the county must, on some level, express the embedded historical violence associated with the ghosting of its Indigenous peoples and question the intangible haunting atmospheres associated with its seductive distance. Now in this time of escalating warming, the white narrative of living in and battling through what always has been a daunting, frightening and isolating space compounded by punishing weather, is becoming unsustainable. Climate change predicts that we are entering a time of transformation. The following discussion acknowledges this transformation and provides a more detailed discussion on Australian Gothic and its links to Climate Gothic. Further, it concedes that Australian Gothic is not just a genre but a state of being claiming a fundamental psychological connection to the Australian places in which we have settled, colonized and now live. The discussion continues by exploring how representation of the country's landscape is in both the physical and psychological response to the current climate-scape.

In contemporary Australian drama, especially that of the late twentieth century and early twenty-first century, the inscribed experiences described above have frequently been interpreted through a Gothicized Australian lens. Prior to a further analysis of the Climate Gothic as a significant emerging genre it is pertinent to open up a broader discussion about what Gothic elements inform it and how this understanding is transformed in an Australian context. Essentially the Australian Gothic situates a ready-made vernacular for revealing the socio-cultural and economic effects of colonizing practices of and in the contemporary Australian world order. Further the Australian Gothic scrutinizes the impact of historical experiences on both geographical and psychological landscapes, and in turn imagines how these experiences impact on the contemporary consciousness. As intimated in the previous chapter, Gothic principles are useful frames of reference for dramatic investigations of climate change as they reveal the inconsistencies of the 'evil' of the phenomenon in the psyche of contemporary Australia. Although Farnell (2014) and other scholars note that the Gothic genre is difficult to define, the term is respected and well established in artistic fields of scholarship. However, it is a particularly contested term especially attracting debate over what the term precisely refers to from contemporary perspectives. In supporting this analysis, it is useful to refer to Goodbody and Johns-Purta's (2019) discussion on cli-fi here as they outline a similar dilemma for literature and explore the quandary from a climate fiction perspective. They posit that climate fiction and literature relating to climate change is not a genre per se as it lacks common plot, formula, circumstance or consistent stylistic conventions that characterize 'genre'. Each novel or story that falls under the umbrella term may be unique and distinctive in its own way. Australian Climate Gothic Drama, like climate fiction, is similarly distinctive and often borrows from and embraces multifarious elements of existing genre attributes. Therefore, while it is difficult to provide a singular definition of either Australian Gothic or Climate Gothic, it is certain that investigation of place, space, landscape, climate-scape and its effects on the Australian psyche are central to the discussion (Hassall, 2014, p. 26). Cohen (1996, pp. 14–16) suggests that from a global position, the Gothic examines the presence of 'evil' as something continuing. Evil or sin as discussed in traditional Gothic modes suggest states of being that are simultaneously real and ethereal and may be difficult to comprehend from contemporary perspectives. Cohen's (1996) reference to such duality, however, seems particularly relevant as he suggests that the contemporary perspective of 'sin' is that which simultaneously identifies the *development of* (actual) and *continuing presence* (imagined) of evil in society. By distinguishing the circumstances and events that are warming the planet as the monstrous or the sinful, there is recognition of a continuing presence that dislocates the relationship between human and non-human nature that legitimately is in need of intervention. The Climate Gothic embraces this sentiment and posits that the spectre of evil lies in the sin of sustaining the political, consumerist and material processes that fuel the climate's voracious appetite for destruction. In analysing the monster that feeds the Climate Gothic vision of Australia, we must first speak to the country's historical beginnings. It is widely acknowledged by Gothic scholars that the isolated land mass at the bottom of an expanding eighteenth-century global map, The Great Southern Land,[1] was originally perceived as an alien space, projecting a terrifying atmosphere, exciting the Romantic imagination as the dungeon at the bottom of the world.[2] European imaginings of Australia as a Gothic space were synonymous with the white imperial experience of transportation and colonization. Gothic sensibilities, therefore, have been embedded in the socio-cultural Australian landscape since Captain James Cook landed on the shores of the East Coast of The Great Southern Land in 1770 and claimed the country for England. Convict transportation occurred concurrently with the popular rise of the Gothic literary genre in Europe. Common sentiment agreed that the most reviled of the English and Irish criminal classes deservedly found themselves transported into an unidentifiable nightmarish landscape claimed through acts of imperial aggression (Carleton, 2012; Gelder, 2012; Hassall, 2014; Turcotte, 1998). I draw on Stephen Carleton's scholarship here to describe how Gothic sensibilities pertaining to estrangement, fear and the need to tame that, which was deemed, uncivilized, resonated in the European consciousness of the time:

> The early generations of inhabitants moving here from the Northern hemisphere felt *estranged* from the Great Southern Land where the hostile indigenous population were seen as extensions of the land itself and needed to be tamed, civilized or erased altogether … (as cited in Hassall, 2014, p. 28)

The historical transplantation of convicts into this frightening continent transformed the death sentence into an experience of surviving a punishing space rife with confusing and monstrous reversals of landscape, place, flora and fauna. For civilized European society, descriptions of Australia were beyond belief. It was an abstract space, a monstrous place that reflected Europe's terrors and phantasmagorical imaginings, resisting all known systems of logic and or reason.[3] The solstices were reversed. The heavenly constellations were upside down. Birds screamed and shrieked. The flowers didn't have a pleasant scent and the plants were dry and brittle—bizarre Grevillea, Banksia, Lantana, mallee and scrub. There was an endless duplication of grey-brown trees that shed bark and kept their leaves. The water was brown and brackish. The native animals lacked logic and were seemingly hybrid freaks. Venomous insects and deadly snakes could paralyze and kill. The waves sported man-eating monsters, while the rivers spawned prehistoric carnivorous reptiles. The land was populated by a naked 'primitive' race that could not speak the language. The weather was extreme, and the climate was untenable.[4] There was too much vast empty distance to become lost in and the never-ending horizon could drive one mad. Anathema to the civilized imagination of European society, the Gothic imagination narrated a penal colony rife with tales of disease and degradation. The deprivation and pain experienced by the First Nation Peoples, the convicts themselves and shortly thereafter, free settlers, set the hallowed ground for the Gothic mindset. The agony experienced by the first Europeans walking on the ancient land is secondary to that of the First Nations Peoples, and their experience of British imperialism intensifies the perception of the Gothic psyche embedded in the contested landscape. The Gothic therefore, is embedded in both the spiritual malaise associated with discovery, transportation[5] and colonization and in the criminality associated with the Nation's beginnings. The Europeans exiled in this punishing space rode with the Four Horsemen of the Apocalypse—drought, erosion, thirst and starvation—and within this paradoxical and dangerous landscape, white Australian settlement was formed. In many respects, it is within these beginnings that the behavior of early convicts and settlers enhanced the monstrous imaginings of Australia and its people. Robert Hughes's (1987) account of this time in *The Fatal Shore: A History of the Transportation of Convicts to Australia 1787–1868*, is firmly embedded in historical accounts and testimonies from transportation to settlement. Revealing acts of cannibalism, bushranger brutality and excessive violence perpetrated on Aboriginal populations, coupled with vicious acts of sodomy, rape and murder rife in a predominantly masculine environment, the forming settlement was further corrupted by the absence of jurisdictive prudence or ethics.

From such contested beginnings, it is evident that Australian history reads like a Gothic horror novel par excellence. When accounting for history as inspiration for the emergence of an Australian Gothic trope, the form reimagines discourses aligned with descriptions of heterotopic spaces, wherein colonies such as Australia are designed as places wherein deviant, criminal or marginalized experiences can be encapsulated (Foucault & Miskowiec, 1986). It was a harsh and terrifying environment, wherein rum was currency, sodomy was rife and punishment was brutal and frequent. The continent's dubious beginnings saw the colonizing population dispersed across the vast primitive landscape, its territory remarkable for the vulgarity of its settlers, the roughness of its few women and the absence of social manners. Playwright and scholar Stephen Carleton (2012) concurs that the grotesque carnival of convict transportation, the subsequent violence associated with the settlement of the land and the deprivation that accompanied Indigenous displacement, abuse and slaughter[6] are signifiers of the social and cultural heritage that has promoted an Australian Gothic sensibility. As a popular form, Australian Gothic drama recognizes the horrors of such an extraordinary history, and further explores how the nation's beginnings informed social life, and in turn, affected the contemporary Australian climate-scape narrative.

Climate Gothic's dramatic and thematic representation extends on the above historical accounts and advocates for a place and space-based critique that includes environmental justice and eco-ethical perspectives. It provides a space for Australian playwrights to question that which has defined and still defines the contemporary culture-scape[7] and allows room for developing a political and ethical inquiry through theatre and drama-based dialogues. As a dramatic discourse, the Climate Gothic specifically responds to material consumerist and economic capitalist behaviours by mutating Gothic patterns of hauntings, monsters and places and spaces into a contemporary eco-critical consciousness. In doing so, it draws sustenance from forces that have impacted on the escalation of warming and climate change in the Australian cultural context. Climate Gothic questions our place in the global world and responds to growing debates about what our white Australian climate responsibilities are. I refer to white Australians here because despite increasing national and global attention about developing essential strategies for Australian climate change adaptation, Indigenous peoples have not been invited to provide their ancient expertise on land management to any major degree. Australian Indigenous leaders have recently[8] reiterated that they have for years predicted a bushfire catastrophe of the scale of the 2019–2020 blazes, which swept the nation for three months. Indigenous leaders are now calling for a radical change to land management. Such criticism describes 'an over-farmed and over-used landscape and suggests a darkly tainted history of colonization as the root of Australia's environmental crisis' (Hassall, 2020, p. 296). By embracing multi-racial and multi-cultural perspectives, the Climate Gothic therefore poses a sophisticated investigation concerning management of land, space and place. Significant to an analysis of Climate Gothic drama scholarship then, is the notion that climate-informed or climate-themed theatre can tease out contemporary climate-scape tensions by posing alternatives to previous twentieth-century definitions of Australian relationships to cultural climate-landscape.

Early Australian settler society very quickly cultivated, farmed and mined the resource rich country. Two hundred and fifty years later our natural landscape, environment and habitat are significantly depleted. In identifying a Gothic tension in the contemporary climate-scape, the landscape becomes a symbol, in many ways a feminine representation,[9] the victim who is brutalized, ravaged and raped. The continued maltreatment of her occurred over a very short period, supporting the (white) quest to build a profitable economy and situate self and nation within the broader world nation space. The metaphoric rape is a narrative motif that resonates through my own investigations into the Climate Gothic[10] and as a dramatic device, it is utilized to expose a careless destruction of habitat, natural non-human environment, non-human species and landscape. The metaphor also reveals the Gothic imagination's preference for power (colonization) over beauty (the natural ecology). The Climate Gothic therefore situates the triumphant 'evil' of the 'rapist-colonizer's' crime of terra nullius[11] and subsequent settler behaviour as that which worked against the well-being of land and habitat. While an offender of violence, the settler/colonizer is in the position of control through experiences that over-awe and ultimately conquer the weaker victim. The Climate Gothic acknowledges the inequity of such processes on the land and how these processes are haunting the contemporary climate condition.

Climate Gothic draws on the power structures of colonizing experiences to articulate contemporary fears about the climate-affected landscape, a landscape that is no longer out there, somewhere in the great distance at the periphery of sight, but is now there outside in our own front yards. If we refer back to Carter's (2010) description of the nation as a stage where history was enacted—a theatrical performance with colonizers as the lead antagonist—we can now broaden that description by suggesting that the playwright (-historian) is now authoring the environmental and ecological legacies we could leave behind. The first performance text analysed in this book is one of my own, *Dust* (2015, 2018). Written as a provocation, *Dust* suggests that our children's children's grandchildren, will look back from the not-so-far distant future and describe the first 25 years of the twentieth century as a time of erosion and extinction. They might describe this Australia as a theatre of dust, a climate-scape wherein our last environmental legacies played out on the nation's stage to an apathetic audience.

Dust (2015, 2018) presents an apocalyptic climate-affected future frontier. Utilizing the materiality of the theatre as a platform to generate environmental conversation, it initially aimed at mobilizing discussion, encouraging social debate and developing strategies about human relationships to, and encounters with, non-human habitats, species and nature. Importantly, *Dust* does not try and solve climate change, but rather, as Hulme (2009) asks of artists, to approach climate change and its ramifications as an imaginative idea in response to our own 'psychological, ethical and spiritual needs' (p. 329). It posed questions such as: what are our personal responsibilities to our land and how are these responsibilities changing in the worsening climate-scape? Uniquely, rather than looking to the past as the Australian Gothic does, the play looks forward and claims the future is that which will haunt, by implying that the future is fixated with the wrongdoings of the history that is being made *now*. The spirit of the catastrophic climate-affected future reveals what we might have done better. The characters—TRIBE—who speak to us from the future want answers about the ecological legacy we have hypothetically left behind to future generations. TRIBE actors, step in and out of character to: (a) pose hypothetical questions to the audience about their climate change responsibilities, and, (b) enact the hypothetical circumstances they are describing. In taking the Climate Gothic position, the play suggests that colonialism is not just a historical event but an ongoing process that continues to have a significant impact on the ecology and environment for future generations. Written as a series of monologues and duologues, *Dust* proposes three time-frames: (1) the days directly preceding a cataclysmic disaster, (2) the years following wherein the country becomes a wasteland and, (3) the far future wherein the extinction of humans as a species is occurring. Predominantly delivered from the perspective of a last surviving community—TRIBE—eking out an existence in the apocalyptic Australian wasteland that is the future, the performance piece explores conflicting positions on the climate change crisis and questions why more didn't happen to avert *The Final Disaster*. Within *Dust* the climate altered space and place is personified[12] and proposes the dysfunctional and affected ecology as the dominant antagonists. Further, dramatic conflict is evoked through human-climate-scape encounters wherein historical, past and present place, habitat, environment, culture and weather take on storied agencies. Significantly, in voicing what has been left unsaid from scientific environmental perspectives—specifically the human and emotional experience of climate change—the play (like all others in this book) extends on research towards the creation of new social and humanitarian imaginaries. Overall, *Dust* utilizes the scientific premise of climate change to frame human conflict in spaces of transition and uses memory of that which is ordinary and familiar as the predominant thematic motif. Further *Dust* reflects and responds to the conflicting and sometimes extraordinary, politicized debates about the environment in which we are living.[13] The excerpt below hypothesizes a future of bleak ecological and environmental legacy exacerbated by political positions of climate denial:

The Prime Minister steps forward and addresses the Australian public. Present time. Speaking from his bunker, 48hrs Before the Day of The Final Disaster.

PRIME MINISTER: We are facing our most, darkest of days. These are difficult times.

Many of you are suffering. Are afraid. However, I urge you not to believe the rumors about worldwide devastation. Propaganda. Yes, in the aftermath of the coming disaster, we will suffer. We have suffered before. But believe me, peoples of this country we will come through this crisis being stronger for it.

LET ME REITERATE!

My environmental decisions have always been firmly based in my total DIS-belief in the severity of climate change. Climate change is a fairytale.

> Global Warming is a fabrication.
>
> I can assure you, no I PROMISE YOU; it will get colder!
>
> Don't believe the hysterical propaganda spread by dissidents, refugees, and terrorists. A lie spread by the Green Left to win votes.
>
> To win votes based on fear. Fears for your and grandchildren's futures. Do not be seduced into believing the lies. I ask you to remember our proud history. A history of beating the odds. Battling anything that is thrown at us, with dignity and with an unshaking belief in God. That is what we do best. We will keep going and rebuild.
>
> Trust in me. Trust in me to lead you through these dark and troubling times and I PROMISE you we will be the stronger for it.
>
> We will fight to protect what is ours. We will fight to KEEP what is ours … I conserve water. I urge you to take shelter. I urge you to protect each other and this great Country. (*Dust*: Yeronga Centre for Performing Arts, Yeronga State High School, Brisbane South. Written and Directed by Linda Hassall, September 2018)

Set in a barren terrain that has been ravaged by climate change and its sycophant, global warming, the play questions if contemporary human society should qualify as an intelligent life form when we take into account what we have done and continue to do to the planet at an alarming rate:

An Australian wasteland 200 years into the future. TRIBE GIRL approaches the audience through the dust.

> TRIBE GIRL: These days are our last days, perhaps? Come Hell or Highwater.
>
> Well …Hell or High Water- is comin'. Yes?
>
> Yes? Yes, it is. So, we is gonna give you a taste. A small taste of Hell and tiny sip of high water. Just in case. Just in case you'se might be interested in seeing what it might be like if you'se continue on. In these disaster days – and you surely might agree's with me that these are disaster days you is livin' – you is livin' and breathin' history. Yes?
>
> Have you thought 'bout that?
>
> You is breathing your last on the edge of the apocalypse in these the days before the Dust. We's is going to introduce you to your legacy – your carnival of dirt. (*Dust*: Metro Arts Production, Sue Benner Theatre, Brisbane City. Written and Directed by Linda Hassall, September 2015)

Dust utilizes Gothic elements within the climate-scape, wherein past, present and future overlap and the characters move in and out of time from some remote point in the future to talk to the audience, to enact the chaos experienced by survivors directly after the disaster where human beings find themselves, surprisingly, on the very edge of extinction:

The Tribal SHAMAN speaks directly to the audience. 200 years into the future. A TRIBE member steps out of character to engage in a hypothetical discussion with another character who is sitting in the audience and who moves into the performance space throughout the scene.

SHAMAN: Before the *Day of the Final Disaster*, in that ghost-speak[14] place, in *your* place…
of time and distance and dust, you'se could not imagine how you and your babes could hunger. We are your legacy and we's know of hunger. Yes?
PERFORMER: Hypothetical.
What happens when there is no food?
A MOTHER: We look for more.
PERFORMER: What if there is no more?
A MOTHER: We go somewhere else.
PERFORMER: What if there is none there?
MOTHER: We go somewhere else.
PERFORMER: How would you get there?
A MOTHER: Drive.
PERFORMER: What if there's no petrol?
A MOTHER: I'd walk.
PERFORMER: To where? Where would you walk to?
A MOTHER: The next town.
PERFORMER: What if there were no more towns?
CHORUS: *(a chant)* Ghost-speak from the Book of Dust.[15]

The time is 8 months after the day of the disaster.

A MOTHER: I walk… I walk… I see the thin skin of my soles scorched into the blistering earth. I feel my little girl. My baby. She is very small. So very pale. Beautiful. So achingly, beautiful. She dies of thirst in my arms. I tear the crusted skin on my arm and chew open my wrist in a vain attempt to nourish my daughter. So beautiful. I tell her I love her to the sun, the moon, the stars and back infinity. I hold her pale beautiful body to me and sing. Sing her name.

I give you a daisy a day dear
I give you a daisy a day
I love you until the rivers run still …

And the four winds have all blown away…

She dies. My Daisy dies and my arms refuse to release her to the dust. So I walk with her. I sing to her and I walk.

I love you until the rivers run still…

I will keep walking. I won't lie down. I will keep hold of her and I will not close my eyes on the dust. I will not sleep. I will not let the dust cover us under a blanket of dirt…

And the four winds have all blown away…[16]

PERFORMER: What would you do if there were nowhere else?

If there was nothing else.

What would you do?

Hypothetically of course. (*Dust*: Metro Arts Production, Sue Benner Theatre, Brisbane City. Written and Directed by Linda Hassall, September 2015)

Dust like much climate fiction, is written as a series of cautionary climate tales—a collection of 'what if' scenarios, based on the premise that if we continue to mine, clear, farm and produce at the current rate and if we continue to ignore the warnings that are rejected in a political arena focused on material and consumerist concerns, we can expect a large-scale environmental catastrophe that predicts our own demise. It frames an environmental and ecological discourse in a world that is now unrecognizable and wherein familiar human experiences in and of the world are transformed:

There are two women, are lost, separately in unfamiliar terrain. Past time. They are unaware of each other's predicament. The LOST GIRL and the LOST WOMAN both have trouble of letting go of that which was familiar. The chorus operate as the lost souls looking for loved ones in the dust and function as both character and commentator. The / indicates where one character's dialogue is cut off and where the other character[s] cuts in. It is 9 days after the day of the disaster.

LOST GIRL: I would like some help. Please/

LOST WOMAN: I put my husband in the shed/

LOST SOULS: John!

Shaun!

Courtney!

Sara!

LOST GIRL: PLEASE! I can't seem to find the place where I grew up. The house with the Big cool verandas and the lemon-scented geraniums/

LOST WOMAN: I didn't want the dogs getting to him. He didn't like dogs/

LOST SOULS: I tried to help
I just wanted to help
Help her
Drag him
She scratched my face she… she… she… bit my/

LOST GIRL: The bus … the bus … the bus … didn't come … PLEASE! I need some/

LOST SOUL: There were people running, screaming into the dust and … and … and … waiting at … at …at places where there used to be bus stops waiting for buses … or asking for/

LOST GIRL: I can't find my house where my parents are … I can't find … PLEASE

LOST WOMAN: I needed to put my husband … my husband in the shed/

LOST SOUL: This woman … this woman … this woman … was dragging her husband by his arm … by the stump of his arm/

LOST WOMAN: He didn't have… didn't have… didn't have… his arm was torn off…I lost his…

LOST SOULS: She was dragging him over the broken bricks and… and… and… and
and… and
Coughing
And Crying
And Chaos/
Jesus Christ! SARAAAA/

LOST GIRL: Sweet Jesus…Help…please…PLEASE… (*Dust*: Metro Arts Production, Sue Benner Theatre, Brisbane City. Written and Directed by Linda Hassall, September 2015).

In its repetition of certain words and/or phrases, its abrupt cutting off of sentences and phrases, the language aims at enhancing the terror felt by the characters in a world where chaos and confusion have replaced that which is familiar. The excerpt is aligned with Trexla and Johns-Purta's (2011) discussion on elements that describe cli-fi literature through detailed positioning of a local future history. The excerpt locates the futurity of a specifically Australian experience, through its inclusion of socio-cultural elements that are not only common, but are loved and familiar: cool verandas and lemon-scented geraniums and further in its socio-masculine preoccupation with the shed as a dominant man space. In doing so, the play promotes an otherworldly experience wherein the familiar is made unheimlich—unfamiliar, strange and frightening. As a

Climate Gothic example, the phenomenon of climate change is positioned, explored and acknowledged as both the theme of the drama and the circumstance of the theatrical world—both the subject and the method of the dramatic conflict. The climate-scape informs the social milieu and the socio-cultural values through the impact of a climate event on the material environment. Further it is that which impedes the characters' ability in achieving their objectives, needs and desires. The climate-scape enhances a sense of displacement and extends on Gothic motifs of absence and loss through haunting imagery. *Dust* therefore contributes to discourse about displacement, memory and loss through its exploration of climate-scape in and of place. Further it questions how all these elements come into play when our beliefs about self in place are significantly disrupted. Like the descriptions of verandas and geraniums and sheds above, the following monologue uses the experience of loss of that which is remembered as a trigger—a 'thing' which triggers both absence and presence. The memory of the thing is present yet the act of remembering evokes a specific psychological association that is recognized as an absence in the climate-scape:

Nora Ellen Nicole is coughing in the dust. She has hold of an old rope suspended from a dead tree. It was once the rope that held a swing. 20 Days after the day of the disaster.

HANGED GIRL: This is...was my grandmother's tree. When I was a child, I would climb up its cool green branches and hide from my grandmother.

She was always wanting me to do things.

CHORUS: NORA...Come inside and make your bed. \
NORA...Come and do the washing up now.
NORA...Go and collect the eggs now Nora.

HANGED GIRL: God what I wouldn't give for an egg. A great big egg from one of my Grandmother's chooks.[17] Fat orange yolk running across one of grandma's bone china plates. Toast dripping with butter cut up into egg-dipping fingers. I won't think of eggs. I won't. I will think of this tree, my grandmother's tree. I will think about how I would hide up in its branches, in the summers when I was young and how in those endless summer days of Christmas holidays its branches would be heavy with fruit/

CHORUS: Mangoes/

HANGED GIRL: Not them Bowen ones. Them stringy mangoes.

CHORUS: Turpentine mangoes/

HANGED GIRL: My grandmother would make chutney out of them turpentine mangoes.

Chutney, full of chunks of ginger and mangoes and swimming in brown sugar and vinegar. I won't think of chutney. Corn meat and chutney sandwiches. I won't think of sandwiches. I will think of the summers when I was young and believed in the magic of this tree and thought my grandmother would be alive forever. In the summer when I was young, I used to think I could live in this tree.

CHORUS: In the summer/
In the summer near Christmas/

HANGED GIRL: In the summer when my grandmother wanted me to do things I didn't want to do. I used to pretend that I could live forever and ever in the tree when the school holidays stretched far into the future. I remember how the mangoes would fall to the ground and rot in the shade of the trees cool green branches.

CHORUS: You mustn't think of mangoes/
Mustn't/

HANGED GIRL: I remember how when I was small there was an old tyre strung from its branches. I recall how my feet would reach higher and higher toward heaven as my grandmother pushed me in the tyre swing. Higher and higher toward the blue sky/

CHORUS: You mustn't think of blue sky/

HANGED GIRL: I won't. I will think about how loud I would shriek/

CHORUS: As you flew toward the sky/

HANGED GIRL: I remember I fell out of the swing and broke my arm. My grandmother rode with me to the hospital and told me/

CHORUS: It'll be all right/

HANGED GIRL: And how when we came home she made me a great big egg with soldiers and gave it to me on a bone china plate. A plate from a bone china setting that someone had given to her for a wedding present.

CHORUS: You must stop thinking of small things/
like mangoes/
and eggs/
and chutney/
and bone china/
and blue skies/

HANGED GIRL: As I got older this tree marked the crying place. It was the place where I came and sat when my grandmother died. I sat under the tree and cried for all the things I didn't do for her.

CHORUS: NORA...Come inside and make your bed/
NORA...Come and do the washing up now/

NORA...Go and collect the eggs now/

HANGED GIRL: For all the help I didn't give her. The rope from the swing is still here. Hanging from the now bare branches of my tree. I focus on the rope. The rope has become everything. It's all I can see. I am too scared to look past the rope. To look past the rope into the distance and the dust. I sit under my grandmother's tree and cry for my grandmother and her big eggs and bone china and chutney.

CHORUS: NORA...Come inside and make your bed.

NORA...Come and do the Washing up now.

NORA...Go and collect the eggs now Nora/

HANGED GIRL: I think about flying toward the sky and wonder if the rope will be strong enough to take my weight. To take my weight when I climb into the tree once again and tie it...tight... around my throat and swing myself out toward the dust. (*Dust*: Metro Arts Production, Sue Benner Theatre, Brisbane City.

Written and Directed by Linda Hassall, September 2015).

Backyard swings have always been a symbol of the Australian dream, representing our healthy, outdoor lifestyle and the symbol that charters a child's journey through long weekends and lengthy school holidays into adulthood. The site is where we sit and dream of the future. The swing narrates our sense of self in place, not only because of its psycho-suburban connotations but also because of its distinction as an object of enjoyment, a symbol of freedom in a welcoming Australian backyard culture-scape. Nora, in recalling long-held perceptions of self in remembered place challenge notions of self in place in present time. Nora's memories of momentous moments of images, tastes, textures, and sounds suggest a lifestyle that no longer exists. Her memories enhance that which is absent and forever lost. The Climate Gothic reading utilizes the act of swinging as a bridge between past and present, a bridge between dreams of backyard landscape and reality of environmentally devoid climate-scape. Once an activity that grounded Nora in self and place, it is now an action facilitating her disappearance into the dust. *Dust* requires us to think about our responsibilities as Australians in a time of climate change, in a time where our sense of place and our connections to home are at risk of becoming absent, lost. It suggests that the familiar is fast becoming unfamiliar and asks us to imagine what might happen to *us*, to our understanding of who we are as individuals and people, when non-human nature is systematically devalued and destroyed.

Notes

1. The name Australia literally means Great South Land.
2. See Stephen Carleton's and Gerry Turcotte's body of scholarly work on Australian Gothic.
3. ibid.
4. There is a wonderful introduction in a book now out of print called *The Australians* by Robert Goodman and George Johnston and originally published by Rigby Limited. In this book the land is discussed in an evocative way and the chapters acknowledge the unique country and its Peoples and the unique cultural consciousness specific to the Australian nation and its characters.
5. See C. Greek's (2017) article on Gothic Criminology which investigates the continued existence of monstrous evil in various modern permutations. Also refer to Picart and Greek's (2010) discussion on colonizing processes as embedding a Gothic criminology. As a narrative feature relevant to this discussion the notion of gothic criminology embraces the phenomenological in colonial and colonizing experiences wherein the vanquisher maintains a god-like rush of power.
6. Carleton further notes that this is compounded with continuing socio-cultural disadvantage of Australian First Nations Peoples.
7. Culture-scape is identified by all or some of the geography and by place, architecture, environment and social groupings.
8. News article by Sydney Morning Herald journalist Patrick Suckling, who discusses '*Why it's so wrong toplaydown Australia's role in fighting climate change.*' (E-News 23 November 2019).
9. See a more detailed discussion on landscape personified as female/victim in Chapter 3.
10. *Dust* (2015 and 2018) and *Salvation* (2013).
11. Terra Nullius is a Latin expression meaning 'nobody's land'. It was a principle used in international law to justify claims that territory may be acquired by a state's occupation of it. British colonization and subsequent Australian land laws were established on the claim that Australia was on discovery terra nullius, justifying acquisition by British occupation without treaty or payment. This effectively denied Indigenous people's prior occupation of and connection to the land and the concept fuels continuing land rights issues in Australia.
12. Personification of other-than human elements is prevalent in the Gothic trope. This idea is discussed more fully in the next chapter.
13. See Marcia McKenzie (2009) "*Scholarship as intervention: Critique, collaboration and the research imagination*", which explores how environmental education researchers might intervene to greater effect in the policies and practices of our governments and local and international organizations through individual and collective programs of research.

14. Ghost-speak is a dramatic term meaning stories that have been carried through the generations from our time (now) to their time (future).
15. The Book of Dust refers to how history as they understand it is written in the Dust of *The Final Disaster*.
16. Song: *I'll Give A Daisy A Day* - Recorded by Strunk, J., Every Little Tune Inc./ Pierre Cossette Music Co., 1972.
17. Chook is an Australian colloquial term for chicken. Many Australian urban homes keep chooks to supply a ready source of fresh eggs.

References

Carleton, S. (2012). Australian gothic theatre and the Northern turn. *Australian Literary Studies, 27*(2), 51–67.

Carter, P. (2010). *The road to Botany Bay: An exploration of landscape and history*. First University of Minnesota Press.

Cohen, J. (1996). *Introduction to monster theory: Reading culture*. University of Minnesota Press.

Farnell, E. M. (2014). Gothic: New directions in media and popular culture. *M/C Journal, 17*(4), 1–6. http://www.journal.media.culture.org.au/index.php/mcjournal/article/viewArticle/880/o

Foucault, M., & Miskowiec, J. (1986). Of other spaces. *Diacritics, 16*(1), 22–27. https://doi.org/10.2307/464648

Gelder, K. (2012). Australian gothic. In D. Punter (Ed.), *A new companion to the gothic* (pp. 379–392). Blackwell Publishing.

Goodbody, A., & Johns-Putra, A. (2019). The rise of the climate change novel. In A. Johns-Putra (Ed.), *Cambridge Critical Concepts: Climate* (pp. 229–245). (Cambridge Critical Concepts). Cambridge University Press.

Greek, C. (2017). Gothic criminology. In A. Brisman, E. Carrabine, & N. South (Eds.), *The Routledge companion to criminology theory and concepts* (pp. 293–296). Routledge.

Hassall, L. (2014). What is Australian gothic theatre? Three playwrights enter the conversation. *NJ Drama Australia Journal, 38*(1), 25–37. https://doi.org/10.1080/14452294.2014.11649570

Hassall, L. (2015). *Dust* produced by Griffith University at the Sue Benner Theatre. Directed by Linda Hassall, September 2015. Metro Arts (unpublished play script).

Hassall, L. (2018, October). Dust produced by Yeronga High School and Griffith University, Brisbane, Directed by Linda Hassall.

Hassall, L. (2020). Contemporary performance and climate change: Redefining the Australian landscape narrative. *ADS, 76*, 270–295.

Hughes, R. (2003). *The fatal shore: A history of the transportation of convicts to Australia 1787–1868*. United Kingdom: Random House.

Hulme, M. (2009). *Why we disagree about climate change: Understanding controversy, inaction and opportunity*. Cambridge University Press.

McKenzie, M. (2009). Scholarship as intervention: Critique, collaboration and the research imagination. *Environmental Education Research, 15*(2), 217–226. https://doi.org/10.1080/13504620902807576.

Trexla, A., & Johns-Purta, A. (2011). Climate change in literature and literary criticism. *Wires Climate Change, 2*(2), 185–200. https://doi.org/10.1002/wcc.105

Turcotte, G. (1998). Australian Gothic. In M. Mulvey-Roberts (Ed.), *The handbook to gothic literature* (pp. 10–19). New York University Press.

CHAPTER 3

The Monstrous Gothic

Abstract The chapter extends discussions about Climate Gothic and its relationship to Australian Gothic through a more in-depth exploration of settler behaviours that have impacted on the environmental health and well-being of the Australian landscape. It transitions the discussion beyond convict transportation to the trauma of the Stolen Generations and further into the future of economy building through resource extraction and impacts. The chapter broadens the discussion on Gothicized elements relevant to Australian and Climate Gothic by identifying historical movements through the landscape linked to various forms of human and other-than-human trauma. Trauma, psychological and physical, is described in terms of the monstrous Gothic and the chapter suggests the monstrous is embedded in the nation-building psyche of Australia. The Climate Gothic discussion draws on experiences of trauma that are associated with the tension between white and Black experiences in colonial landscape and present climate-scape. The play discussed, *Salvation* (Hassall, L. (2013). Salvation. TEXT Special Issue Website Series: Scriptwriting as creative writing research, 19, 1–21. Retrieved from textjournal.com.au/speciss/issue19/Hassall.pdf), investigates how diasporic spatial experiences associated with European and Indigenous relationships to place are crucial to the Climate Gothic perspective and, further, how these experiences have had long-term impacts on the well-being of the Australian natural environment.

L. Hassall, *Theatres of Dust*,
https://doi.org/10.1007/978-981-16-6159-4_3

Keywords Convicts · Race · Settlement

The monstrous as discussed in this chapter refers to the impact of colonial behaviour and associated traumas on the contemporary climate-scape. Gothic associations of sacrilege, slaughter, sacrifice, defilement and decay evoke pertinent monstrous images and while they are words associated with the Gothic genre, they can also be used to describe the effects of climate change and the escalating destruction of land and habitat on a warming planet. Unlike the monstrous embodiments explored in popular American Southern Gothic traditions rife with vampires and supernatural beings, Australian Gothic is often secular; its ghosts and demons are all too human (Carleton, 2012). The Climate Gothic has been alluded to previously as embedded in convict settlement, early settler behaviour and associated diasporic experiences of European and Indigenous populations. The following theatrical discussion moves past convict transplantation through to the Stolen Generations of aboriginal children removed from their families and placed in white 'care' and further into a future of resource extraction to consider how colonial disjunctions of space and place are crucial in analysing the current climate-scape from an Australian perspective. What was unimaginable yesterday is shockingly real today as the appearance of climate change destroys our previously held beliefs about the natural Australian landscape. From the contemporary perspective, climate change may be understood as the appearance of the contemporary monster of our own making and when discussed in relationship to Australia as a Gothic space, can be described as the contemporary atrocity on par with any Penny Dreadful man-made horror.

The monster's body in the contemporary climate can be identified as a cultural body, and as Cohen (1996) so eloquently suggests, "it serves as a harbinger of category crisis that renders porous categories of knowledge and form and makes fluid the realms of fear [;] it guards the borders of the possible [and] comes to dwell among us" (pp. 12, 13). One of the most striking features of considering such an evocative overview as pertinent here, is that the Australian Gothic consciousness has always tended to prey upon itself and to savour, consume and salvage persistent convict and settler motifs that are now feeding into the Climate Gothic. Climate change and its grotesque paramour global warming dwell silently among us, until in voracious rage they raze communities and feed on already deteriorating landscapes. Their fury-driven feasts are becoming more and more frequent as they unleash their weather minions, which maintain a remarkably frequent presence within the cultural milieu. Their presence within the Climate Gothic manifests as a sign of community omen, an auger of trauma which provides material for Australian playwrights to interpret through the dramatic lens. Residing in the twilight realm between fact and disbelief, climate change's monstrous destructive scale stimulates the imagination to inform the emergence of the Climate Gothic.

As outlined in the first chapter, the Australian Gothic argues the settler foundational narratives of brave new worlds to expose the haunting traumas shaped by colonial life. The sins of the buried past resurface and disturb the present, whereby Australian socio-cultural positions about race, place, nation and people are exposed and questioned. The historical landscape, with its many terrifying phenomena that were unnameable in the English language, impacted on the colonial experience; at the same time, the colonial settler experience braved that which was unidentifiable in an attempt to tame the landscape into an uncivilized and uncultivated, yet desired, little Britain (Carleton, 2012, Turcotte, 1998). The conversion of transportation into colonization and the subsequent social settlements that formed, saw the transplantation of (white) social and cultural organizations throughout the heart of the landscape (Carter, 2010). The atmospheric phenomena embedded in the landscape coupled with aggressive settler behaviours inform the twentieth-century's Australian Gothic investigations into colonial culture and its repercussions, specifically in the area of race relations, which prescribes the continuing politicized Indigenous discourse about ongoing trauma and displacement.[1]

The convict, described previously as a Gothic symbol of crime and punishment, comes to us from the pages of oppressive Australian history. I draw here on Carter's (2010) scholarship which identifies a convict as a 'social and cultural construction [existing] as a reflection of a body of rules, as a personification of transgression, a figure of speech necessary to the ruling class's self-justification and the perpetuation of its power' (p. 295). The convict is acknowledged as a symbol of imperial aggression—one who was silenced by structures of judicial punishment and imperial power (p. 295). The play discussed in this chapter *Salvation* places the convict and colonizing experience in a larger 'framework of transnational, multi-cultural and imperial world history', offering what Curthoys (2008, p. 147) describes as a 'glimpse into the large-scale systems of governance, surveillance and power' as was situated within British imperial influence. Imperial influences which impacted on the Australian climate-scape can be attributed to colonial beginnings based on transportation. The experience of convict transportation to the New World coupled with diasporic experience and the punishing incarcerated circumstances exist to justify the appropriation of the historical space.[2] Such experiences map the beginnings of the misappropriation and mismanagement of land, place and space, and narrate the foundations of ecological pressures stemming from colonizing processes of farming and mining. In the ecological situatedness in contemporary Australia, there is, on one level, a sense of 'horrification' at ongoing development, mining and farming industry practices that raze, excavate, farm and clear. However, the mining industry monster employs 1.9% of Australia's work force. Herein lies the paradox. Australian natural resources such as coal and gas in addition to meat export, historically have kept the economy afloat, being the lifeblood of the nation's fiscal security. In the contemporary COVID culture-scape, Australians are politically aligned with a system that in many ways ignores the impact of these industries on the landscape, as employment stability and economic growth are at the forefront of the nation's consciousness.[3] Therefore, amidst Australia's significant climate change crisis, the debate concerning resource extraction continues. These statistics suggest that the mining boom is far from over in this country, yet heated debates over the proposed ADANI mining expansion in Queensland continue. Australians remain divided over what they perceive as the threat *to* the job security that mining provides many Australians or the threat *of* the mining industry and its causal effects on the natural habitat. However, on a positive note, the name ADANI is now a rallying cry for a growing community of activists, and their ecological protests are being heard across the Nation's political stage.

National tension concerning the colonization of the land and consequent nation-building is ongoing, and the conflicts incorporated through debates pertaining to race and land entitlement premised another of my original performance texts, *Salvation* (Hassall, 2013). *Salvation*, a precursor to *Dust*, explores how the spread of white culture escalated the disappearance of Black culture over the last 250 years and in turn, exacerbated the effects of climate change in this country. It specifically investigates themes of (white) cultural and environmental legacies in and of the Australian landscape. Articulated as a contemporary performance piece, the play explores a chorus of Gothic-inspired ethereal characters who are identified as overlapping VOICES haunting the PAST, PRESENT and FUTURE Australia. The deliberate overlapping of timeframes suggests that the past not only haunts the present but that both past and present will haunt the future. This structure is utilized as a dramatic device that suggests time is malleable as is synonymous with Gothic fiction. Importantly, the ever-changing spatio-time continuum experiments with the way contemporary drama can narrate and debate the white colonial story by overcoming realistic elements of the outer world, e.g., space, time and causation. Consistent with style and theme in traditional Gothic fiction, the *Salvation* world excerpts an almost dream-like atmosphere which enables the Nation's repressed thoughts and actions to surface. Exploring the repercussions of the birthing and dying of the Australian Nation, *Salvation* offers an alternative theatricalized version of Australian history. It does so through a series of surrealistic and nightmarish fragments depicting monstrous imperial activity including transportation, land appropriation, the systematic mistreatment of First Nations Peoples and the development and proliferation of industries which have impacted on the environment. In identifying the play as an example of Climate Gothic, it is pertinent to go back to our historical beginnings before moving forward to analyse specific historical Gothic elements that are embedded in the European obsession with appropriating, owning, dividing and mismanaging land.

In the opening moments of *Salvation*, the monstrous symbol of transportation—the TALL SHIP—is personified and sets the tone for the encompassing Gothic atmosphere that informs the climate-scape throughout the play:

(*A TALL SHIP is breathing, coughing and gagging. A cry of pain. A moan of release. Through the mist and the distance, a bell is tolling. Convicts are spewed from the bowels of the ship – they evoke the voices from the past*).

PAST: I am one of the dead, for surely one cannot survive this. Yes, dead I must be. Yet in death there seems no comfort. From the endless sea. The endless sickness.

CHORUS: Singing Too-ra-la-oo-ra-la -addity! Singing Too-ra-la-oo-ra-la-ay! Singing Too-ra-la-oo-ra-la -addity!

And we're bound for Botany Bay[4] …

PAST: With every rancid breath I taste excrement. Shit festering with disease. It is not a taste easily spat out. It is the taste of misery. A collective filth. The taste of starvation. A scurvied flavor lingers on the tongue. We stand on the bloated bellies of the dead. Our scabbed and putrid feet sink into rotting flesh. Our toenails turn black and peel away. The rats are well fed and thrive.

PAST: Men heave against each other. Sweat against each other. Sodomise each other. Or are sodomised. Men.

PAST: Men stronger than I summon the strength to defy the monstrous sea, this monstrous demon ship, this monstrous vengeful journey to the dungeon of the world. We go for God and England and Justice. Men find comfort in the little death of release. Men defy social sensibilities to feel something. Anything. Rave and Rape against the injustices of England's transplanted Justice. We journey as aliens into the unknown. God help me. Destination Botany Bay.

CHORUS: Singing Tooral liooral liaddity Singing Tooral liooral liay (continues)

(*Convicts are sucked back into the TALL SHIP. As it approaches the sounds of human misery magnify*) (Hassall, 2013, p. 3).

The TALL SHIP scene theatricalizes the moment before European invasion and its Gothicity is embedded in the disjunction between the cruel experience of transportation and the hopeful anticipation of a future settlement. The TALL SHIP with its billowing *white* sails stretching towards the beach is the symbol of the 'evil' of convict transportation and also for the spread of Europeans into and through the landscape. It is personified as the beast, that in coming to the end of its laborious journey, hatches the first depraved and diseased white populace onto the ancient shore. As such it aligns with the Gothic fascination of personification. The TALL SHIP situates the colonial prison experience within a (post) contemporary Gothic theatricality. The ship is the overarching scenographic construct, e.g., the representation of a white Australian Nation about to come into being. Its creaking timbers expose the framework of both the dramatic landscape and the theatrical culture-scape that is experienced by the transportees. It is both a symbol of imprisonment and a symbol of colonization. Being entrapped in the monstrous belly of the TALL SHIP enhances the claustrophobic terror of its actual dark imprisoning space, while at the same time the experience of transportation enhances the psychological trauma associated with estrangement from place as associated with the Australian Gothic trope. From either a literal or symbolic viewpoint, the TALL SHIP may be viewed as a vessel of forced migration, the method of actioning punishment, that which wrenched people from country or place of origin to transplant them into foreign soil that purportedly was uninhabited. From these moments, the theme of colonization in the play describes an abuse of power, perpetrated by the colonizer's desire to settle and tame the country, by their craving to build their own community and establish civilized society which would create the thriving resource-based economy that has significantly contributed to the Australian climate crisis.

Embedded within the play is a sense of eco-historicism described by Hiltner (2009) as that which contemplates climate-influenced human culture from historical positions and is directly linked to European empire-building, longstanding racism and to the disturbing ecological lessons we are now privy to (Trexlar & Johns-Purta, 2011). From this perspective, *Salvation* gives voice to some uncomfortable national silences to provoke socio-cultural questions pertaining to identity, nation and race relations that have indirectly impacted on the contemporary environmental situation. The play acknowledges Hartz's (1964) theories that suggest Australia is frozen within its moments of formation, and offers an alternative dramatic reading of this moment from a (post)colonial spatial perspective. It questions historical Eurocentric landscape discourses, asking 'What is the environmental legacy we will leave behind to future generations if we continue on in the same way?' In doing so, *Salvation* aims to unsettle ideological observations about nation-building that have promoted popular myths about European dominance and domestication in and over the Australian landscape.

The *Salvation* culture-scape excerpts a form of psychological violence over the human characters, who find themselves estranged in PAST, PRESENT and FUTURE places and spaces that are unfamiliar and who respond, in turn, by overlaying violent behaviours into environment, place and space. From the moment the TALL SHIP disgorges its occupants on the Eastern shore of The Great Southern Land, the embedded Gothic perspective delights in reversing the idealization of the domestic or the ***heimlich*** (familiar, homely) by making all things familiar strange or unheimlich. The Gothic duality of its heimlich-unheimlich atmosphere aligns with Freud's much discussed theories, which suggest that the unheimlich becomes truly uncanny when it collapses into its reversal, when the heimlich becomes strange. In *Salvation*, the settler experience of colonization mutates the heimlich, wherein nostalgia and a desire for that which is familiar are altered experiences enhanced by the unheimlich Gothic space. The dual experience disrupts human attempts to conjure or replicate any sense of home or familiar comfort. The heimlich becomes something foreign, something frightening in a frightening spatial experience of endless, unforgiving, brutal and condemning distance. Contrary to passive concepts of place, the settler experience embodies the space with something other than passive or comforting. It transforms domestic sensibilities into haunting psychological experiences of a space that resists domesticating. McClintock (1995) suggests that the term domesticating, when used in the active sense as described here, is akin to the meaning of dominating and from this perspective acts of or attempts to domesticate the landscape prescribe the action of civilizing. The Gothic paradox is in the liminal psychological space that sits between domesticating and dominating. This paradox is strongly evoked in *Salvation*.

The dramatic tension between the heimlich and the unheimlich sits within memories and recollections of sounds, smells and belongings that evoke familiar comforts. The sensory and tactile memories sit securely in spaces other than the ones experienced as disjointed and disruptive[5] and which reject any conjuring of joy. In the following excerpt, nostalgic (mis)remembrances of desired 'other' places and spaces are explored from the character of FUTURE. This perspective generates a Climate Gothic, almost poeticized atmosphere resonating with foreboding and a surreal sense of melancholy for what is lost. In attempting to reveal the moments directly preceding settlement the scene offers an eerie ambience, echoing with the sounds and smells of a past place—now a space of perceived salvation that in reality doesn't exist, and may never:

As the settlers move through the landscape, the experiences of colonization, rather than offering salvation for a new beginning, foster ruptured unheimlich experiences.

FUTURE: The empty vessels of settlement carried our language – never before uttered, never before heard – yet the noise of it shattered the silence and shredded the distance.

FUTURE: Language. You brought the English language into a landscape that could have easily done without it.

FUTURE: You babbled into the distance and told it of buildings we planned to construct.

FUTURE: You told it inconsequential things.

FUTURE: You complained about its heat.

FUTURE: You told it about curtain materials and rugs and crockery and babies yet to be born.

FUTURE: You named places in the distance, paddocks and mountains and other things that had no use for your useless names.

FUTURE: In building and cursing and debating and naming you settled your own history and mapped out a future.

FUTURE: The distance remained silent. Her silence echoed her contempt (Hassall, 2013, p. 7).[6]

The acute desire for that which evokes nostalgia becomes active, a behaviour or an encounter which offers history disguised as fiction. The series of common domestic life experiences and the anticipation of building home, society and nation created the monsters responsible for the unheimlich climate-scapes in which we now live. A sense of nostalgia for a place that welcomes remembered domestic trivialities is prevalent. The dramatic gaze back to the past can only reveal that which is painful in the contemporary milieu. In turn, memory is linked to intellectual, social, political, economic and environmental histories within the liminality of time, memory and place where the fate of the colonizer-now-victim is determined.

In many ways the scene (continued below) reads as a metaphor for nation-forming, a metaphor that weaves together continuing environmental, racial and cultural tensions:

PAST: Salvation was thought to be found in the great distance. In our desire to move into the emptiness of that vast desert space.

PAST: To spread ourselves, however thinly across – into – through the horizon.

PAST: We left white boot-heeled footprints in the dirt.

PAST: We trod carelessly across the darker bare footed imprints of those of ancient lines and eternal time.

PAST: We didn't care. We were seeking Salvation.

PAST: We took little with us.

PAST: Nothing of import really except God and Justice and Good intentions.

PAST: We carried these with us on the backs of dreams of land.

PAST: Dreams carried in cardboard suitcases and boxes and buckets.

PAST: Vessels filled with goods and greed and great white hopes.

PAST: You see, Salvation was ours sayeth the Lord.

PAST: In Salvation there was promise. The promised land.

(*Time is malleable and the past blends into the future. The FUTURE voices are male*).

FUTURE: Never mind the genocide.
FUTURE: Never mind that.
FUTURE: That's a position of perspective.
FUTURE: History.
FUTURE: You were just moving into the distance eager to meet the promise.
FUTURE: The promise of Salvation from the past.
FUTURE: The promise of Salvation from poverty and the pain of separation.
PAST: We looked into the distance and imagined beyond into the brightest of bright futures.
FUTURE: You could not conceive of what you did…what you caused…what you ignored…what you…what you-
FUTURE: -what you see now from far into the future you couldn't possibly have imagined.
FUTURE: You were like children (Hassall, 2013, p. 5).

Curthoys' (2008) research alerts us to the fact that as a nation we have steered away from discussing themes such as some raised in *Salvation*. Curthoys claims, for example, that it is only recently that historians have begun to "consider the word genocide as applying to Australian history" (148). Coupled with themes that touch on the mass slaughter of Aboriginal populations, *Salvation* explores Indigenous trauma as being linked to diasporic experiences including how Aboriginal minority populations were forcibly removed into places of resettlement which left them traumatized and homeless, fringe dwellers in their own ancient land. In thematically addressing the forcible removal of children, the Stolen Generations,[7] and utilising expressionist physical theatre forms, this shameful and bleak aspect of history is woven through the theatrical experiences explored as forging the Australian frontier:

PREACHER: Cast your eyes across this godforsaken land and look for him. Search until you are blind and dumb and deaf. But do not give up. He is out there in the drought and distance. Yes, he is you, miserable sinners. Salvation is yours sayeth the Lord.
CHORUS: At the cross, at the cross, There's salvation full and free at the cross[8]…
PREACHER: Find the heathen and bring him to the cross.

(Aboriginal women and children are dragged to the cross and treated brutally).

PREACHER: For his own good and your salvation bring him to the cross. If he refuses bind him. If he still refuses to cast his eyes toward the lord, toward the true cross of God, flog him until his back is bleeding and broken.

CHORUS: There thy soul shall find sweet rest, At the cross; Heaven's peace shall fill thy breast, All thy guilt shall pass away, At the cross ...

PREACHER: It is your responsibility, designated to YOU from God to terrify him, terrorize him and if that doesn't work TEMPT HIM with disease riddled blankets and glass beads into believing. Take his women, take his country then take his spirit. And if that does not convince the misguided heathen to turn his eyes toward the wooden cross, kill him and take his children.

(The children are torn from their mother's arms and given to the white missionaries waiting at the cross).

The scene frames a monstrous Gothic theatricality incorporating religious fanaticism, ideological conversion of the 'other' and in its incorporation of a missionary zeal and brutality, the scene examples what Hoge and Mishra (1991) describe as an historical obsession with the guilt associated with the forcible removal of Aboriginal children. In highlighting the religious, missionary imagination, the monstrous activities of colonial and (post)colonial missionary practices provoke questions pertaining to the fearsome temporal and spiritual forces operative in the haunted landscape of history. Again, positing a reversal of (un)truths associated with the white historical narrative, *Salvation* brings to the fore questions about trauma experienced in a landscape wherein religious fanaticism is anathema to any experience of social justice or salvation. The theatrical climate-scape, therefore, includes the liminal space between myth and (white) history and describes a space where white colonial characters' inhibitions are minimized, and the scope of monstrous consciousness broadens. In linking Gothic themes of forcible removal of children and mistreatment of First Nation communities with the Climate Gothic trope, *Salvation* depicts a paradoxical world, one that stands in opposition to common, rational or moral experiences narrated about *being in* and the *well-being of* the landscape. The scene also premises how the treatment of First Nations Australian Peoples has had significant repercussions on their identity and on their culture and cultural legacy.

By incorporating debates that expose what is hidden about the past, the play responds to experiences of possession and dispossession. Embedding these themes in FUTURE viewpoints, invasion-occupation matters are explored through a contemporary lens. First Nations Australian Peoples' relationship with country (as opposed to landscape) and the country's well-being 'is complex and multi-faceted [and] intricately linked to caring for the country and environment' (Ganesharajah, 2009, p. 6). Deborah Rose's (1996) much cited and evocative definition of this relationship states:

> Country is multi-dimensional – it consists of people, animals, plants, Dreamings, underground, earth, soils, minerals, water, air ... People talk about country in the same way they would talk about a person: they speak to country, sing to country, feel sorry for country and long for country. (p. 7)

Rose describes country as a natural being, one that deserves respect, a living entity that should be nurtured in a harmonious and reciprocal relationship. European settlement, however, brought to Australia a totally new fundamental idea to Indigenous experience—landscape as a devisable and alienable commodity. This system of commodification directly aligns with Climate Gothic exploration relating to the white notion that land can be owned to satisfy our desire for possession of space and domesticating of place. The anti-colonial critique of white imperialist activity aligns with Plumwood's (2008) analysis of how white Australia grew from an insensitive sense of self in place. However, place appropriation for profit is typical to colonialism. It claimed that land could be *owned* by an individual rather than nurtured by a community (Forbes & Spearitt, 2003, p. 25). From a Climate Gothic analytical perspective, this sensibility therefore inspires dramatic examination on the impact of colonizing actions on environment and place:

> PAST: She called to me. I moved toward her. By the good Christ she was a beauty. Strong and fierce. Bright and hot. Her name – Horizon. She captivated me with the colors she clothed herself in. Shades of seduction that would never be recaptured in paint or oil or imagination. She whispered words on the hot red winds that blew from the West. Words that made me weak. That made me tremble. I desired all of her. Every inch of sandy hot hills and pink tipped crests. I climbed over, through and into her. I had to have her, tame her, beat her into submission and then she'd be mine. I sinned to have her. By the Christ I did. I was lost in her distance. (Hassall, 2013, p. 7)

Within the contradiction between salvation and damnation and in the battling of the weather and the distance, lies the proliferation of the white Nation's developing economic structures and the resulting impact of over-resource extraction and land exploitation. In its examination of the quest for profit, *Salvation* explores how early capitalism embraced imperialism to seize Australia's prolific natural resources. In doing so it questions contemporary economically driven activities from the perspective of the FUTURE and warns that there is only so much that can be extracted or taken from the land before (she) stops giving:

> FUTURE: Why were anyone here in the first place I'd like ta know.
> FUTURE: When?
> FUTURE: Before the weather. (*A dingo howls in the empty distance*). Before the bastard dogs and bleak weather. Before one fuckin' weather disaster after another. Before we got buried in Salvation.
> FUTURE: Holocaust initially. That was then forgotten, rewritten. Then it was the Coal or the Copper or the Cattle. Gas or Oil. Must'a been some fucking good reason ta come out here after all that went on. All that violence. All that murderin'. How the fuck would I know? Men came. That's what I know. Men came and attempted to make a mark.
> FUTURE: She's [the land] opening up. Out there. Cracking and splitting-
> VOICE: Gets us all 'ventually. The distance.
> VOICE: It's difficult to move through anymore. She's opening up. Out there. Cracking and splitting – (Hassall, 2013, p. 17)

An analysis of *Salvation's* psychological climate-scape provides a Climate Gothic perspective on the mechanisms of cultural systems such as mining and its links and flows of capital, social exclusion and class stratification. Imperial sèizure of land for resource extraction is described by Lyman (1990, 1992) in true contemporary Gothic mode, as ***vampiric consumerism*** (my italics). It could be suggested that this term is very apt when we consider how parasitical consumerism supported natural resource extraction for profit in the shaping of the Nation. Ironically, consumerist practices relied on the colonizer/aggressor's ability to exploit Indigenous Australians to work in these industries. The vampiric nature of economic growth through resource extraction is understood as being at the expense of the natural landscape and on its Native people who were in spiritual and environmental harmony with the land. In what is arguably the most variable and unseasonable climate on earth, First

Nations Peoples of Australia have lived successfully and more than sustainably for 3500 years and survived many extreme climate change events (Robin & Smith, 2007). In the ***Dark side of the dream: Australian literature and the postcolonial mind***, Hodge and Mishra (1991) discuss at length how Europeans preferred imposing imperial practices on the landscape for the purpose of creating a flexible set of ways of programming "a nexus of rights and obligations toward the land" (p. 92). Such rights of land ownership allowed the infiltrating white settlers to bring their stock and further practice excessive demands on a fragile ecology that the country and its First Nations People had never before encountered. Spence (1866, historic account as cited in Johnston & Anderson, 2005) describes the successful colonial economic perspective. I refer to Spence's (1866) observation to provide further clarity on the exploitation of the land and on its original owners:

> … we have an enormous territory sparsely populated by an agricultural, pastoral and mining population … and here and there a township close to a gold field or a copper mine … we only produce raw material and all our efforts are directed toward producing it with the smallest amount of labor … In Australia, a great proportion of the land is unappropriated and held by Government in trust for the [white] people … I am so much accustomed to look forward a few generations for the future of our own colony and wherever I turn I see so many instances of the economy of land, so many proofs of its enormous money [sic] value, so much care taken of it, and of all that can be supposed to increase its productive powers, that it is impossible for me to overlook the greatest of all distinctions between the new country and the old. (in Johnston & Anderson, 2005, pp. 50–56)

As Spence outlines, until the early 1900s, white Australia was a conflicted nation caught between the monstrous desire to kill off the entire Aboriginal population and the requirements of their vampiric pastoral needs to use them as an essential 'slave' labour force (Johnston & Anderson, 2005; Biskup, 1973; McQueen, 2004). As such, an almost classical colonial situation developed; political and economic power was in the hands of the Europeans, while the Aboriginal communities retained the unwelcome status of terra nullius. The continuing fiscal activities of resource extraction and the expansion of pastoral needs saw immense economic growth in the fledgling white nation. However, these factors are now being attributed as supporting climate change's extreme weather conditions, including drought, flood, fire and wind erosion. In what may now be described as a Climate Gothic reversal of fortune, the aggressor has become the victim. Industry reliant Australians find themselves in a liminal state, that of remaining within the almost inescapable man-made structures they have developed and perpetrated within place, while simultaneously being in and beyond it due to climate change escalation. In the present dysfunctional political economy, the liminal state generates a Climate Gothic critique, wherein a state of fear is layered into contemporary society's overall playing out of exasperation, aggression, rejection and denial about how carbon emissions are contributing to worsening global warming. The historical mutation of the natural landscape examples the proliferation of human spaces reliant on material culture. The mutation is evident in a system of values and standards that are layered into the transition from landscape to climate-scape. Europeans arrived and, with little forethought to perpetuating a sustainable landscape, they began to cut down trees, farm that which was previously unfarmed, exploit the land for resources and transplant models of architecture and culture. *Salvation* (2013) presents the exploitation of non-human habitat as ongoing:

(Men are moving roughly, loudly, drunkenly in and through the landscape. The men are engaged in numerous physical labours that move through time. They are claiming the land as their own. They are determined to tame it. It includes tree felling, sawing, hammering, cane cutting, mining—any industry that has raped the landscape for natural resources. The industrial soundscape is created by the actors with the tin buckets. The landscape bleeds. The men tire—collapse and proceed to drink themselves stupid).

> PAST: Letter Home 5: Dear Temperance You have been long suffering in the face of my absence. This bastard place binds me to cruel labor. My back is near crippled, my hands are blistered and weeping. No matter how hard I work there seems more to be done. The sun drives me mad. It puts me in mind of murder. (Hassall, 2013, pp. 16–17)

Thematically PAST's Letter Home aligns with Gothic sensibilities from the twentieth century, whereby white displacement is enhanced by extreme weather, a dramatic construct prevalent in Climate Gothic wherein notions of displacement are presented through the pervasive representations of environmental degradation. Gothic paradoxes are further embedded in eco-historical perspectives, which claim that the climate significantly influenced the Nation's collective psychology:

PAST: My Dearest Faith We are surely damned. There is no Salvation to be found here. Even if we repent. There is drought. There is terrifying distance …

CHORUS: Terror Australis (terror) Terror Nullius (terror). (Hassall, 2013, p. 5).

The ruin of place and space is a final condemnation of the white colonial experience:

> FUTURE: Someone else desires her. I could not keep her. Not forever. Fool that I was. I loved her too cruelly. Too long. She did not forgive me. Her colors have faded. In her sadness her former brightness flickers. Her generous sandy hills are left scarred. Deep pits of emptiness where her beauty used to be. I took from her all she had to give. She rejects me. And my heart breaks. My heart is torn from my chest … The red winds that blow from the West no longer whisper but scream my betrayal. She hurls her drought and fire and flood at me and scorns my fear. She wants another. A lover who will be tender. In the dark eyes of the ancient lover I see my pale defeat. (Hassall, 2013, p. 16)

The scenes suggest that any hope for an ecologically healthy FUTURE environment has been defeated by actions and choices made in the PAST, and which continue to be made in the PRESENT. Colonizing and (post)colonizing practices have offended the ancient landscape. Climate change as the monstrous beast of our own making is destroying the mythic settler landscape that we have proudly narrated to the world. It is using extreme and catastrophic weather events as its defense mechanism against those who have taken all she had to give. Gendering the landscape as 'she', a brutalized female space, aligns with contemporary post-feminist Gothic investigations whereby manifestations of inherent violence are working through the nooks and crannies of the psychological and physical environment. *Salvation* situates a feminist eco-critical analysis within the aesthetics and personification of place. The metamorphosis of non-human nature into a gendered representation as described in the excerpt above, is similar to Gothic fiction popular in the Romantic period. Climate Gothic analysis in *Salvation* points out the limitations and significantly, the effects of masculine modes of thought and behaviour that have traditionally been enacted and reported as history. The inversion of the once rich and fertile ancient landscape of pre-white history is now a barren, infertile wasteland, which is noted by the voices of the PRESENT in the play:

PRESENT: This landscape is a woman going through menopause. Once fertile, her rich red soil is no longer fertile. She is a hostile and cruel bitch, my landscape, as she cracks and no longer bleeds, opening herself up for the rain that doesn't come. (p. 13)

As part of the anti-colonial critique, the eco-feminist reading extends the examination and or understanding of the contemporary trauma associated with the climate-scape. Supported by traditional Gothic sensibilities, various traumatic states are examined through a feminized Climate Gothic lens. The lens reflects trauma themes through multiple imaginings, perspectives and illusions about contemporary settler cultures and the maladjustment of Eurocentric persons and practices in the land which they appropriated and settled.[9] The analysis of *Salvation* suggests climate change is a malleable thematic idea and can be framed as dramatic discourse that is both political and ideological. As Hulme (2009) explains, it is a container of multiple stories, narrated through a range of discourses and includes rhetoric, political biases, philosophical beliefs and individual interpretation. *Salvation* examines sites of multiple stories and discourses from PAST, PRESENT and FUTURE positions to provide a dramatic hypothesis on trauma, time, displacement, memory and loss, and how these experiences have in some ways interpreted the contemporary atmospheric climate-scape.

Notes

1. See Stephen Carleton's award-winning play, *Constance Drinkwater and the Final Days of Somerset* (2006). The play investigates the desire to civilize the landscape and touches on the abuse of the Indigenous population in the quest to do so. The relentless weather drives everyone mad in the end.
2. See Carter's (2010) extended discussion on this theme in the *Road To Botany Bay*.
3. In 2018, the Resource Minster revealed that Australian mining exports boosted the economy by $174 billion with the Labor Market Information Portal claiming that the mining industry employs approximately 233 300 people.
4. *Bound For Botany Bay* is a well known Ballad sung in Australia and it is referenced in various documentaries and films researching the transportation of convicts to the continent, although the practice of transportation to NSW ended in 1840, 45 years before the song was actually written.

5. See Urban's (2007) discussion concerning disruption and nostalgia in *Ghosts from an Imperfect Place* which explores Philip Ridley's three major dramatic works from the 1990s. The article serves as powerful reminder of nostalgia's origin as a physical affliction caused by an acute longing for home and the past.
6. This passage is particularly inspired by Paul Carter's extraordinary work, *The Road to Botany Bay: An exploration of landscape and history*. The book, a radical account of Australia's spatial history, dissects imperialism and explores how Australia was formed through acts of settlement, possession and dispossession. See Chapter 5: Debatable Land, where Carter discusses how the continent was linguistically settled.
7. The Stolen Generations are the children of Aboriginal and Torres Straight Islander descendants who were forcibly removed from their families by Australian Federal and State Government agencies and by church missions as a result of various government policies between 1890–1970. The impact of these experiences has generated generational trauma and grief associated with loss of connection to family, identity, language, culture and country.
8. Hymn: *At the Cross* by William J. Henry 1911.
9. See Deborah Bird Rose, (2002) *Dialogues with Place: Toward an ecological body*, which she explains, is a decolonizing endeavor to explore 'place' as a partner in landscape dialogues.

References

Biskup, P. (1973). *Not slaves, not citizens: The aboriginal problem in Western Australia, 1898–1954*. University of Queensland Press.

Carleton, S. (2012). Australian gothic theatre and the northern turn. *Australian Literary Studies, 27*(2), 51–66.

Carter, P. (2010). *The road to Botany Bay: An exploration of landscape and history*. First University of Minnesota Press.

Cohen, J. (1996). *Introduction to monster theory: Reading culture*. University of Minnesota Press.

Curthoys, A. (2008). Does Australian history have a future? *Australian Historical Studies, 33*(118), 140–152. https://doi.org/10.1080/10314610208596187

Forbes, J., & Spearitt, P. (2003). Rum corps to white-shoe brigade. *Griffith Review, 2*, 23–35.

Ganesharajah, C. (2009). Indigenous health and well being. *Native Title Research Report No. 1/209 AIATSIS, April*, pp. 1–43.

Hassall, L. (2013). Salvation. *TEXT Special Issue Website Series: Scriptwriting as creative writing research, 19*, 1–21. textjournal.com.au/speciss/issue19/Hassall.pdf.

Hartz, L. (1964). *The founding of new societies: Studies in the history of the United States, Latin America, South Africa, Canada and Australia*. Harcourt, Brace and World Publishers.

Hiltner, K. (2009). Renaissance literature and our contemporary attitude toward global warming. *Interdisciplinary Studies Literature and Environment, 16*(3), 429–441.

Hodge, R. I. V., & Mishra, V. (1991). *Dark side of the dream: Australian literature and the postcolonial mind*. Allen & Unwin.

Hulme, M. (2009). *Why we disagree about climate change: Understanding controversy, inaction and opportunity*. Cambridge University Press.

Lyman, S. M. (1990). Rereading Robert E. Park: Toward a gothic perspective on capitalism and imperialism. In M. W. duQuesnay (Ed.), *Explorations: The age of enlightenment* (pp. 29–108). University of Southwestern Louisiana Foundation.

McClintock, A. (1995). *Imperial leather: Race, gender and sexuality in the colonial context*. Routledge.

McQueen, H. (2004). *Social sketches of Australia, 1888–2001*. University of Queensland Press.

Lyman, S. M. (1992). *Militarism, imperialism and racial accommodation: An analysis and interpretation of the early writings of Robert E. Park*. University of Arkansas Press.

Plumwood, V. (2008). Shadow places and the politics of dwelling. *Australian Humanities Review, 44*, 139–150. http://australianhumanitiesreview.org/2008/03/01/shadow-places-and-the-politics-of-dwelling/

Robin, L., & Smith, M. (2007). Science and place in time: Archeology, ecology and environmental history. In C. Dickman, D. Lunney, & S. Burgin (Eds), *Animals of arid Australia: Out on their own?* (pp. 188–196). Mosman: Royal Zoological Society of New South Wales. https://doi.org/10.7882/FS.2007.049

Rose, D. (1996). *Nourishing terrains: Australian Aboriginal views of landscape and wilderness*. Australian Heritage Commission.

Rose, D. B. (2002). Dialogues with place: Toward an ecological body. *Journal of Narrative Theory, 32*(3), 311–325.

Spence. (1866/2005). In J. Johnston & M. Anderson, *Australia imagined: Views from the British periodical press 1800–1900* (pp. 50–56). University of Western Australia Press.

Trexla, A., & Johns-Purta, A. (2011). Climate change in literature and literary criticism. *Wires Climate Change, 2*(2), 185–200. https://doi.org/10.1002/wcc.105

Turcotte, G. (1998). Australian Gothic. In M. Mulvey-Roberts (Ed.), *The handbook to gothic literature* (pp. 10–19). New York University Press. http://ro.unow.edu.au.cgi/viewcontent.cgi?article=1060&content=artspapers

Urban, K. (2007). Ghosts from an imperfect place: Phillip Ridley's nostalgia. *Modern Drama, 50*(3), 325–345.

CHAPTER 4

Gothic Atmosphere and Extinction

Abstract Atmosphere is discussed in this chapter as one of the most evocative elements of the Climate Gothic mode as it works in relationship between the historical landscape and the historical manifestations of recent history to inform the contemporary climate-scape. Theatrical atmospheres position climate, temperature and often brutal weather, in association with isolating psychological experiences. Within this experience, environmental criticism including species extinction can be located. In the Climate Gothic plays discussed in this chapter, the dramatists are preoccupied with species survival and set the works in the deep and darkly atmospheric Tasmanian wilderness. In true Gothic style the playwrights utilize the paradox of that which is already extinct to investigate that which is on the brink of extinction. This chapter discusses two plays; *They Saw a Thylacine* (2017) and *The Dead Devils of Cockle Creek* (2018) wherein the brutally atmospheric texts expose the ideological distance between us as a species and the natural and environmental habitats affecting non-human nature.

Keywords Extinction · Tasmanian Gothic · Eco-feminism

In exploring extinction as a theme in recent Australian Climate Gothic drama, our Gothic sensibilities require us to once again return to history,

L. Hassall, *Theatres of Dust*,
https://doi.org/10.1007/978-981-16-6159-4_4

to the time in 1770 when Captain James Cook first sailed into the Bay of what is now Sydney Harbor, to map the East Coast of Australia. Cook and his company must have found themselves in a true Gothic duality of experience—that of admiring the beauty of the Sydney Cove landscape and being over-awed by what they saw. Perhaps they were astonished by, even fearful of, the bizarre wildlife. Further they may have been alarmed at the silent presence of Aboriginal peoples—perhaps members from the Borogegal, Birrabirrigal, Cammeraygal, Gadigal, Gayamagal, Wallumedegal, Wangal and Gadigal people, witnessing the arrival. These experiences still have the power to stimulate the artistic imagination. We envision lush bushland forest fringing the beach. We imagine how it must have teemed with unique Australian marsupials such as quolls, koalas, kangaroos, wombats and bandicoots. The skies would have been rich with rainbow hued birdlife. Approximately 250 years later, there is a species extinction crisis in the country with many of the once abundant native species in significant decline. Ecologists describe Australia as a megadiverse country, yet one that is boasting a sullied reputation of leading the world in mammalian extinctions and having the highest number of threatened species on the planet (Hughes, 2021; Robin, 2010). Australia is widely documented as having one of the worst extinction rates of any nation, yet environmental research posits that approximately 8 million hectares of threatened species habitat were cleared between 2000 and 2017, suggesting Australia's Environmental Protection and Biodiversity Conservation Act (1999) is 'ineffective at protecting potential habitat for terrestrial threatened species, terrestrial migratory species or threatened ecological communities' (Ward et al., 2019, p. 1). There have been approximately 100 species listed as extinct in Australia since colonization, with 1790 more species listed as threatened with extinction in 2019 (Foley, 2020; Ward et al., 2019). Two of our cherished, unique and iconic animals are on the list. The koala population is significantly endangered and has halved in the last 20 years. Kangaroos are being hunted in the largest commercial slaughter of land-based wildlife on the planet. Almost 90 million kangaroos and wallabies were lawfully killed for commercial purposes between the years 2000 and 2020. Ironically, more than a million kangaroos were killed in Australia in 2020 alone as the country launched a mass cull of the creatures to protect endangered grasslands and environment. Additionally, invasive species of flora and fauna pose a greater risk for native habitats and wildlife. The spread of invasive weeds across the continent, especially since the 2019–20 bushfire catastrophe which razed the country for three months, is "replacing endemic species, dominating eco-systems and reducing plant diversity" (Foley, 2020, n.p.). These unprecedented bushfires ravaged an area approximately three times the size of Tasmania, killing or displacing at least three billion native animals.[1] In true Gothic mode, the laws of non-human nature are being significantly challenged in a country that is already described as the most arid on the planet.

Extinction of our native species can be traced back to the 1930s with the last Tasmanian Tiger dying in captivity at the Hobart Zoo in Tasmania in 1936. Tasmania is described by Bullock (2017) as Australia's haunting geographic equivalent to literary explorations in the popular swamp riddled, but simultaneously humidly seductive, American Southern Gothic trope. Tasmania, the Great Southern Land's smallest island state is by contrast starkly beautiful; its cold and often bleak weather conditions prescribe long periods of icy isolation for its many smaller communities. With a total land mass of just over 68,400 km, it is separated from the southern end of country by the treacherous winds and waves of the Bass Straight and lapped by the freezing waters of the Southern Ocean. Its very disconnection from the larger land mass of the country has promoted inaccurate but popular derogatory Gothic descriptions of a space where in-bred grotesques cohabitate. Contemporary mainland opinion satirizes the population as maintaining its limited gene pools from throwbacks to escaped convicts and cannibalistic bushrangers. Identified as the crucible for the Nation's past horrors and unsettled past, Carleton (2014) discusses the tiny island state as an atmospheric mutation of European Gothic through its infamy of being the site of the country's most brutal penal settlement, Port Arthur. He discusses how fear driven from both inner and outer sources 'flourish in [the Tasmanian] space – a space that dictated confinement not only in prison cells but on the island colony itself' (p. 29). From a Gothic perspective, the Port Arthur space is a mutation of the haunted castle architecture of the traditional Gothic mode and in some ways symbolizes the imposing structure of the Imperialist convict judicial system whereby the specter of the Empire casts a large shadow across the tiny island state. Adding to its identity as one of our foremost Gothic zones, Tasmania's Port Arthur penal colony mutated into that of the country's and the world's most horrific massacre[2] sites in recent times, when Martin Bryant committed what was, at the time, identified as the world's largest shooting by a lone gunman. The incident left 35 people dead and 18 wounded. The atmospheric space reverberates strongly within the Australian psyche and throughout the country at large, projecting a monstrous atmosphere that strongly evokes the Gothic sensibility in its distinctive weaving together of past and present manifestations of violence in both penal outpost and mass slaughter site.[3]

This is the atmospheric space which frames and, in many ways, informs the thematic conflicts in the two plays discussed in this chapter. Both are set in the Tasmanian wilderness and both explore the extinction of species, specifically the Thylacine and its close relative the Tasmanian Devil. *They Saw a Thylacine* (2017) by Justine Campbell and Sarah Hamilton investigates the extinction of the Thylacine, popularly referred to as the Tasmanian Tiger. Excessive hunting combined with factors such as captivity and habitat destruction led to the rapid extinction of the Thylacine in the 1930s, which numbered 5000 at European settlement. Through dual descriptions of the last days of two Thylacines, the subtextual undertones reflect the colonial profiteering practices of bounty hunting into present concerns regarding the captivity and extinction of species. Incorporating a soundscape that is eerily Tasmanian Gothic par excellence, the haunting cries of animals and birds are juxtaposed with the sound of imprisonment: cages locking, chains clinking and creaking resonating across and through the cries of urban captivity and zoo administration. The Zoo is modelled on the haunted patriarchal spaces of imperial Gothic environments, a space of abuse and neglect and a place that embodies a dogged determination to civilize the barbaric wilderness of history. The aural superimposing of the symbolic past into the depressing present, infuses the work with a melancholic atmosphere, and similar to the evocation of landscapes in the developing catalog of disturbing Tasmanian Gothic films (Bullock, 2017), it generates a haunting claustrophobic and frightening *mise en scene*. *The Dead Devils of Cockle Creek* (2018) by Kathryn Marquet explores the decimation of the Thylacine's closest relative, the Tasmanian Devil. While the Devil's plight unfolds, the spectre of the last Thylacine in captivity haunts the isolated ranger station in which the play is set. Both the Thylacine and the almost extinct Devil project mythic bestial presence in the folklore of the Nation, and symbolically prescribe notions of danger and of things buried deep within the Tasmanian wilderness. The Thylacine symbolizes species extinction in the Australian context (Varney, 2015) and both plays in various ways evoke its loss as both a unique animal and a folkloric shadow presence, haunting the edges of the contemporary imagination.

Both plays embed our desire for non-human species to continue to exist through and beyond what Alaimo (2019) discusses as the Sixth Great extinction. Hamilton and Campbell set *They Saw a Thylacine* (2017) in an evocatively eerie Tasmania of the 1930s. The play navigates the death of the last Thylacines. Extinction themes are investigated through the brutal and careless treatment of the animals. The female characters articulate their awe of the last surviving animals and the care that should be taken of them in what is predominantly a masculinized environment.

They Saw a Thylacine describes the decline and extinction of two animals. Through two interweaving monologues, the story of two animals' last days is recounted by BEATIE McCULLOCH, a regional, country woman and tracker of the animals, and ALISON REID[4] an aspiring zookeeper. The two women bear witness to the last days of the Thylacines, one in the wild and one at the Hobart Beaumaris Zoo in the 1930s. Framing the extinction account is the possibility of a third animal, a vague conjectural presence, a sighting of a shape, a shadow, a stalking specter of a beast that is seen to disappear into the Tasmanian wild and whose footprints are tracked as heading into the present, haunting the contemporary milieu with the Australian extinction crisis. The play is an example of what Varney (2015) describes as a combination of ecology and theatrical engagement in its proposing that species extinction is a major concern for contemporary governments, sciences, cultural arts making and broader humanities and social sciences.

This play is Climate Gothic par excellence and brings to the fore the violence of cruel and atmospheric Tasmanian weather conditions in which the women and the Tigers must experience. Evoking a freezing atmospheric climate-scape, non-human nature is violently personified, e.g., fire *spits* and *whips* at BEATTIE and *tells* her quick-witted yarns (Campbell & Hamilton, 2017, p. 5), the trees *whack* at her head and ice *pounds* and *whips* at her cheeks (Campbell & Hamilton, 2017, p. 10). In weathering the climate's slights against her person, she watches as the sun *bleeds* into Deadman's creek (Campbell & Hamilton, 2017, p. 9). As she tracks the elusive Tiger, BEATTIE tells us that dawn wakes her with a *slap* and reminds her that there is a Thylacine waiting to be captured:

A TALL SHADOW

BEATTIE Break o'day wakes me with a slap.

> Not a foul slap – the slap says 'Congratulations! You are alive! Well done!
> *What a bloody blighter! Have courage! There's a Thylacine out there*
> *for you and ain't that bloody grand!* (Campbell & Hamilton, 2017, p. 15)

The play's rich and evocative language generates a hybrid Gothic vision, combining the colloquialism of the Australian vernacular with the stylistic to enhance the geographic and natural elements that are imbued with the power to hurt and/or disrupt. In personifying weather and nature's role in encouraging the hunt, the play explores the paradoxical dialectics of place and species (*dis*) placement.

Processes of extinction are recounted through a narrative of animal sightings, their capture, their escape and finally through the animals' eventual death. Their disappearance out of existence is experienced first-hand by BEATTIE and ALISON. In *They Saw a Thylacine*, the Climate Gothic paradoxes emerge through the contrasting behaviours of BEATTIE's hunting of the animal in the uncivilized and threatening wild, e.g.:

> BEATTIE: You leadin me further further
> And further
> The whistling wind persistent in my ear now
> Curls round my chin like a cold snake
> Chokin me for fear
> Nuzzlin deep, hopin for strength to
> Throttle me good
> And that's when I hear it
> Yip yip yip
> Hear it
> On the wind
> Not far though I'm thinkin, not far
> And there it is
> Double the sound
> Yip and then YIP and then YIP and then YIP
> It's all my might I use not to YIP back and shout
> *STAY WHERE YOU ARE, TIGRESS BEAUTY BEAST THYLACINE,*
> I'M COMIN AND WE'LL BE DANDY FINE YOU KNOW! (pp. 12–13).

…and in ALISON's concern for Ben (the incongruous name of the female Thylacine) held in 'civilized' captivity at the Hobart Zoo:

> ALLISON: … I start to hear it regular
> The yip yip yap
> Of Ben in the dark
> Meaning she's been locked out of her den
> Got nowhere to park
> Herself away from the cold (Campbell & Hamilton, 2017, pp. 28–29).

The separate viewpoints constructed around the YIPPING of the animal in the free wild and in the captive zoo environment heighten thematic dialectics of place and displacement. Displacement from both spaces is evident (a) in the Thylacine's need to escape from hunters in the wild and (b) from neglectful captors in the zoo. The tension between the two places brings into prominence questions arising about the value of species protection premised by captivity. Ironically, one of the most significant Gothic paradoxes in the work occurs when BEATTIE herself is captured, having fallen into a Thylacine trap set by male hunters. When BEATTIE gains her freedom from the 'high sided', 'man-cut' wood, 'hatched trap' (pp. 13–14), she imitates the Thylacine's call, vocalizing her human ability to beat the hunt:

> BEATTIE: I grab rope from the rucksack and throw it
> Throw it
> Throw it through the thatch overhead!
> I hang from it, pullin with both me arms and makin myself heavy as possible
> And fucken heave down on the man-made roof which has me covered in everythin
> Hold breath, hold breath tight and close eyes sharp and pound pound pound the
> Head and kick kick kick like the wind and her cold snake
> HOT ROPE PULLS MY HAND TILL IT BURNS
> I've got to hold on for dear –
> And down comes the stupid trap roof till I can see my way out,
> climbin walls for sheer joy and escape is my victory – YIP YIP
> YIPPPPPEEEEEEEEEE!!!!!!!
> I'm not beat and eat
> Some possum

For sheer delight. (p. 14)

The dramatic irony of this scene is evident. Once captured, there was no escape for the Thylacine from a trap such as this; nor would the beast be able to celebrate with a possum dinner caught in the wild—possum being one of the Thylacine's staple food sources.

A significant Gothic paradox in the work comes into play with the Tiger's escape from the Zoo. Escape suggests that it will never be dominated or held captive to human species commodification; however it will only be free in death, its extinct reality haunting the Island. Its act of survival and triumph involves a denial of capture and captivity processes that inevitably leads to its extinction. The Climate Gothic can be read in the play's mediation between the polarities of the natural habitat, which is now a site of human intervention and capture, and in the containment and administrative patterns of the Zoo's unnatural environment. The duality of experiences recounted depicts the struggle to reconcile the split division of two worlds – natural and colonized. Within this division, the characters offer differing accounts of nature and culture through BEATTIE's hunting experiences in the wild and ALLISON's exclusion from and critique of urban zoo processes. It is through these elements the playwrights draw on elements of the female Gothic:

ALLISON: ...But while at the zoo
Dad taught me all he knew
Though in the nineteen thirties
Knowledge wasn't the only thing to get you through
See I couldn't pick up where Dad left off
Because of my tits
Tiny as they are
They wouldn't take me as far as I needed to go
And I needed to go into the zoo
But I didn't have a cock so I didn't get a key
So I didn't get a key
And I think that's what stopped me
Which in turn led to a chain of events
That saw the demise of a fair few beasts
Including Ben
But back then who knew
Our last living Thylacine
Captive Tassie Tiger
Would cease to exist. (p. 12)

ALLISON is advised she will not be employed at the zoo because the Board of male directors who administer the zoo environment do not want to be criticized for nepotism (her father was the previous administrator). She tells us that 'different rules apply to those with a cock... than those with a cunt' (p. 19). ALLISON herself is often referred to as 'pet' by males in the middle management positions and as the sexist economic and social conditions play out, the fragile existence of the unique species is exacerbated. ALLISON, like the oppressed TIGER, rejects her subservient role and subjugated position in the Zoo. An eco-feminist reading is also evident in BEATTIE's awe of the 'Tige' in its natural habitat, even while she is set on hunting and capturing it. She croons to it, and calls into the dense night for the 'Crafty Tige', the Tige that has 'Mystic powers ... Like dark nights gaze' (p. 10). While she is in some ways fearful of the beast, she ultimately respects its almost mysterious qualities in its natural environment. However, she has few qualms about her role in its possible capture and confinement: "Well I'm gonna take you to the zoo, Tige, where you'll be fed n full n famous" (p. 11). For BEATTIE, like her father before her and the generations of men before him, the Tige loses its identity as a species and becomes an economic symbol, an object of capitalist profit:

> BEATTIE: I follow the old snoz.[5]
> Got it from the Ol Man[6] and I won't give up on it now
> You'll come again with a winking head and tail flick
> You'll say, '*Don't worry, I'm with her*'
> An I'll say, '*Lay down your guns, we mean no harm*
> *I'm her to collect my prize*
> *The money, all of it, thank you very much sir*
> *Yes, that's right sir, I brought you what you sought*
> *Yes, that's right sir, I'm a woman and I'm wearing a skirt*
> (Campbell & Hamilton, 2017, p. 11)

The excerpt suggests a feminization of the dominant masculine role in the Tasmanian bounty-hunter space. BEATTIE's excessive immersion in the Tasmanian wilds and her ability to hunt, capture and make money from doing so, jeopardizes the patriarchal order of the bounty-hunter role. The excerpts suggest an oscillation between victim and victimizer while the extinction process of hunting, capturing and imprisoning, controls the lives of all female characters including Ben.

Overall, the dramatic climate-scape investigates the consequence of human actions whereby different interest groups represented by BEATTIE and ALLISON are trying to shape spaces and species according to their own needs and desires. The tragedy of the Gothic duality relates to the Thylacine's need to escape the hunters and captivity to live *and* the need for it to remain in captivity to stay alive.

The Dead Devils of Cockle Creek by Kathryn Marquet (2018), also set in the deep and dislocating Tasmanian wilderness, in many ways investigates a future wherein beliefs concerning the reconciliation between human and non-human species is no longer possible. Set in a small shack in the wilds of South West Tasmania, GEORGE is a female environmental scientist who is trying to save the Tasmanian Devil from extinction in a time of climate change. Thwarted by HARRIS, a ranger employed to cull the feral dog population, and two unexpected visitors, MICKEY O'TOOLE, a fast food magnate and DESTINEE, a lost girl guide, GEORGE in her devotion to the Devil population, sets off a series of violent events. Framing the thematic exploration is the proliferation of disease within the Devil population, which is escalating the threat of their imminent extinction. However, the unique Gothic elements of this work lie in its incorporation of an absurdist duality, i.e., the disease in the last surviving Devil population juxtaposed against the disease of unquenchable consumerism within contemporary society. The psychological atmosphere also incorporates dual perspectives. Concepts relating to non-human nature and beauty are denigrated while simultaneously the rotten and offensive aspects of human nature are applauded. Demons of folklore are revised, as capitalist markets of consumption are viewed through the deal the character MICKEY makes with HARRIS, who is a fan and consumer of the product. MICKEY is an Irishman whose fortune lies in unethical practices related to the profitable concern of chicken nugget manufacturing. Exploding traditional assumptions, customs and beliefs concerning fair trade and marketing, MICKEY's 'sin' is exposed in the horrific make-up of a mass-produced product that exploits the diseased Devils. We are horrified to learn that MICKEY's company is harvesting the Devils to use their cancerous tumours as flavour enhancement in the product. The play, like *They Saw a Thylacine*, disturbs notions of identity and culture through a Gothic exploration of dismaying experiences of human dominance that represent the new world order. Indeed, the escalation of contemporary Gothic themes align with the socio-cultural and economic impacts of globalization wherein people and material products and consumer developments transition with increasing speed across national and international boundaries (Byron, 2012).

What is most disturbing about this work lies in Marquet's take-no-prisoners investigation of the dying out of the species through the unregulated gluttony of species commodification. There is no doubt whatsoever that the Tasmanian Devil is on the brink of annihilation due to human intervention. After HARRIS discovers what he believes is a corpse in the bathroom of the shack, we learn that GEORGE has shot MICKEY when she discovers him brutally killing the Devils in their pen:

> GEORGE: The alarm tripped and I'm thinking it's ferals. And, I saw your rifle there and I just – grabbed it. And, I was practically running down the ridge towards the creek and then – then – there was that dickhead, standing right by the den, hacking into my devils. There's only one left. *[she indicates the pouch she is wearing]*. And, I couldn't see, Harris. I couldn't fucking – I just pulled the trigger. (p. 22)

GEORGE explains to MICKEY, who has revived but is still seriously injured, that the Devils he was so brutally slaughtering are the first Devils born free of disease in generations, signifying some hope for species continuation:

> GEORGE: Um, thing is Mickey – the work I do here is very important. Um, as you are probably aware, the Tasmanian Devils are in crisis because of this cancer that's so contagious and it's just decimated the population to the point where it's unlikely that the species will recover. And, I've been studying them for nearly ten years and it's all been bad news. But then, you see, one day last year, I found a family – down at Cockle Creek – and they'd come into contact with the cancer virus, but they didn't catch it. They couldn't catch it. There were no symptoms. See? (p. 36)

MICKEY's careless slaughter of the Devils resides, like in *They Saw a Thylacine,* with the reflection of the ghosts of animals past who were hunted, captured and killed for profit. Further extinction processes lie in inflections of colonial belief systems that ignored the value of native species. As a Climate Gothic example, the play problematizes capitalism's version of reality and its relationship with species, environment and habitat through its thematic assault on domestic consumption of a popular branded frozen food source. In doing so *The Dead Devils of*

Cockle Creek gothicizes the traumatic relationship between knowledge of species extinction and human indifference to the extinction problem. The absurdist and satiric form Marquet employs reveals the failure of political and environmental structures that perceive climate change, species extinction and global warming as distant impersonal threats:

> GEORGE: ...We took the stars out of the sky and fashioned ourselves as gods. And if I talked until my voice was hoarse, it wouldn't make a scrap of difference because I won't change your mind, and that is exhausting [...] Okay, we know – not suspect – not guess – we know that we are causing rapid climate change...It isn't a debatable issue – with two logical and rational sides – it's a fact and we're already in the midst of it. But stupid people or politicians with coal succubus in their wallets, walk around sprouting "How do we know?" Fucking Jesus ... (p. 52)

The play aligns potential species extinction with environmental concerns that examine the arrogance of human behaviour through the human characters, most notably through the character of DESTINEE. The teenage girl guide becomes symbolic of an Australian demographic who lack or ignore environmental foresight. Through her we understand that we must eventually, despite apathy and argument, reckon with our staggering impact on the planet and our fellow creatures.[7] DESTINEE's right-wing rhetoric aligns with growing Australian political discourses, specifically the growth and popularity of Pauline Hanson's One Nation party and its nationalist and conservative agenda:

> DESTINEE: I'll stop terrorism and make heaps of jobs for Australians and cut environmental red tape and not let any mussies [Muslims] or Chinese people come here. I'm going to be like Donald Trump and make Australia great again (p. 69).

The One Nation party primarily supports a white One Nation and popularly supports the continuance of dominant white identity by defending white sovereignty over the landscape and what we do with it. Ironically, DESTINEE is a second-generation Asian Australian,[8] and her beliefs are firmly aligned with those of her parents who embrace the uninformed ideological zeal spouted by the far-right wing political party, believing that assimilating this sort of rhetoric is what makes them true-blue Aussies. The scene entitled EPISTOMOLOGY (pp. 65–73) sees DESTINEE become a mouthpiece for environmental misconceptions. The realization of DESTINEE as a clueless antagonist in the play in many ways supports Alaimo's theories (2019, p. 398) that question if it is at all possible to disconnect contemporary political epistemologies of global environmentalism from propagated colonial histories and practices:

DESTINEE: The planet will be fine. Climate Change is a big fat lie.
GEORGE: Sorry?
DESTINEE: If global warming's a thing, why is it snowing here for the first time in ages?
GEORGE: It's much more complicated / than that.
DESTINEE: Bae, it's a conspiracy. Scientists. NASA. China. They're all in on it. (p. 66)

And a little later in the scene she adds further fuel to the fire by spouting popular right wing religious rhetoric in her support of continued resource extraction:

DESTINEE: You're just gullible and believe all the lies. If we don't use the resources god gave us, we're throwing them back in his face. (p. 67)

Ironically, the *destiny* of the continuation of the Tasmanian Devil species is left in the hands of DESTINEE after a 'culling' of the three other characters occurs at the climax of the play. MICKEY brutally murders both GEORGE and HARRIS and stumbles into the Tasmanian darkness, while whistling the jingle to O'Toole's Premium Quality Nuggets. He then succumbs (we suspect) to the gunshot wound he sustained at the beginning of the play:

DESTINEE clings to the pouch with the baby Tasmanian Devil. Behind her the monitors beep and the computers come to life [the security system]. MICKEY walks past one of the cameras, stops and falls, unconscious to the ground. DESTINEE looks at the baby in the pouch. She touches it.

DESTINEE: It's alright. I got you.

A flash of the Thylacine pacing in its enclosure.

Lights.

The end.

The Thylacine pacing in its enclosure, sighted at the end of *The Dead Devils of Cockle Creek* is Ben the animal protagonist depicted in *They Saw a Thylacine*. The intertextuality of the animal's appearance is accidental but in *The Dead Devils of Cockle Creek* appearing 80 years later, it projects an uneasy, unheimlich sense, a Gothic disruption of time, space and place. In *The Dead Devils of Cockle Creek* Ben's menacing shadow presence pacing in her enclosure, overwhelms the climate-scape. We know she will pace until her eventual escape and death. After death the symbolic spirit of her extinction is the vague conjectural presence, the sighting and shadow of the once magnificent and unique beast that disappears from this play, only to reappear and disappear into the Tasmanian wilds of the past in *They Saw a Thylacine*. Her footprints tracked as heading into the present and back into the past, remind us of other species who have no future, and Ben symbolizes all animals on the brink of extinction. The haunting presence of the Thylacine in both plays is the Australian equivalent of what Varney (2015, p. 6) discusses as speciesism, a term that frames the willful neglect of animals and which identifies the basis of our different treatment of animals and humans.[9]

As intimated earlier, what is significant about both plays is that they incorporate what can be identified as Female Gothic and eco-feminist principles. Specifically, second wave feminism acknowledges that Gothicism is intrinsic in lives of women who have been subject to patriarchal socio-cultural conditioning (Armitt, 2007). Patriarchal conditioning is especially evident in the relationship the human and non-human females have to the environment wherein a fundamental cruelty works its way through the nooks and crannies of the climate-scapes depicted. The climate-scapes as embedded in both plays, become a site of female repression, both non-human and human, a space where the female women and Tiger(s) and Devils are cast off, neglected, collected, displayed, abused and tossed aside like commodities. The atmospheric freezing and haunted wilderness is the home to the extinct Tiger and the almost extinct Devil, and the human interventionist habitats of the Hobart Beaumaris Zoo in the 1930s and the nugget manufacturing plant in the twenty-first century, control to some extent the lives of the human and non-human characters. The human and non-human females are victims of colonial processes. The spaces and places in which they interact cast them adrift in oppressive and inescapable male-dominated power structures. As such, both plays encapsulate eco-*femme*-centred perspectives, which see the characters dealing with untenable hierarchal male behaviours that have disrupted the non-human species equilibrium:

ALLISON: You'd think you don't have to go have gone to Yale
To know the difference 'tween

> A tiger with a penis
> And a tiger with a pouch
> But there it is
> They didn't know
> The men who captured her
> They named her Benjamin
> We called her Ben
> Soon as I saw her I knew (Campbell & Hamilton, 2017, *They Saw a Thylacine*, p. 11).

And …

> GEORGE: Oh let's forget what he [MICKEY] did. What he stands for. How we spend our lives fighting against people like him.
>
> …
>
> Devils can't speak our language. They can't defend themselves. If we don't stand up for them, we're allowing genocide to continue. Why won't you hear me?
>
> HARRIS: Because it's too hard. Its too fucking hard. You're too hard. Why can't we try and help the planet without adding to the misery. Why can't we have balance.
>
> You're the only person in this revolution. And you won't let anyone else in. No ones good enough for you. So if you can't give just an inch, give me a scotch, valium and cocaine. I'll settle in and watch the end of the world off my face (Marquet, 2018, *The Dead Devils of Cockle Creek*, pp. 82–83).

GEORGE is symbolic of the traditional Gothic 'governess', her charges the last of the Devil species, her environmental uniform symbolizing both her belief in species protection and her moral governance concerning a disappearing species. MICKEY symbolizes a barbaric modern civilization which is explored further in HARRIS's apathetic attitude to the exploitation of place and species. The Climate Gothic frames the Female Gothic elements through representations of species and characters that oscillate around issues of victimization, responsibility and autonomy. As examples of Climate Gothic, each work incorporates an intersection of the Gothic and feminism through an exploration of speciesism and through depictions of the inherent violence that works through manifestations of the psycho-atmospheric Tasmanian landscape:

GEORGE: So you have to understand my shock and horror to find that this very group of devils who could have been the saviors of the entire species were being massacred – by you – right in front , my eyes. You can understand – why I pointed the gun at you and shot you (Marquet, 2018, *The Dead Devils of Cockle Creek,* p. 36).

The female characters in both plays attempt to negotiate their relationship with species through interactions with space, place, culture and environments that are, from a paradoxical Gothic viewpoint, explored through gendered-environmental roles.

The plays are significant to the development of Climate Gothic analysis as they confront inequities of non-human extinction issues and ask for greater ethical treatment of species to address the escalation of species in crisis.

The Dead Devils of Cockle Creek uses human dramatic conflict to make an eco-critical statement about our impact on the natural habitat, while *They Saw a Thylacine* gives precedence to an eco-critical account of species extinction over the human drama. However, both plays provide us with contemporary narratives that explore species extinction as part of the rupture of the natural environment. Both plays propose the themes of non-human species inheritance and disappearance which highlight the (post)colonial crisis of extinction by identifying the parts we played in its escalation. In the Anthropocene we have been consumed with orienting towards the human presence on the planet. The plays discussed in this chapter suggest we should pay more attention to caring for the presence of the other. They alert us to the disappearing tracks of what is now the mythic non-human species treading into extinction across the changing natural environment that was once considered as the epic Australian landscape. The following chapter further explores the intrinsic relationship between the people and the places that defined popular Australian myths developed and perpetuated since colonization. Even as species were disappearing before his eyes, the mythic relationship with the land largely supports the white man's right to a space he tamed and fought for.

Notes

1. See Lesley Hughes's (2021) discussion on the postnatural, post human and postwild in *Griffith Review's Issue 71* Remaking the Balance. Hughes focuses on humanity's troubled relationship with the Earth and identifies

how humans have significantly transformed the Earth in a very short space of time.

2. The **massacre** refers to the mass shooting in and around Port Arthur, Tasmania, on April 28–29, 1996 by Martin Bryant. At the time it was described as the largest massacre by a single gunman in the world.
3. Playwright Tom Holloway (2007) in *Beyond the Neck* provides us with a haunting and moving account of the Port Arthur massacre ten years after the event. A verbatim piece collated from testimonies, the play explores the effects of long-lasting trauma on an isolated Tasmanian community. The play is a reflection on the terrifying atmospheric elements emanating from Tasmania and poses a Gothic theatricality that suggests the distance between the atrocities of the colonial past are layered within the contemporary landscape. While not concerned with climate change per se, it does provoke discussion about (post)colonial experiences of specific Australian sites, such as Tasmania, that are populating the Climate Gothic trope.
4. An eccentric figure in 1930s Tasmania Hobart's society, Allison Reid for many years assisted her father in the management of the Beaumaris Zoo until his death in 1935. While she was acknowledged as playing a critical role in the running of the Zoo during her father's curatorship, she was passed over as his successor in what is described as an appalling demonstration of misogyny. See J. Dunlevie's *Curious Hobart: What Happened to The Beaumaris Zoo?* ABC New Blog on-line 10 July 2018.
5. Colloquialism for nose.
6. Colloquialism for father.
7. See Alaimo's (2019) discussion on environmental aesthetics where she questions if it is possible to disconnect epistemologies, politics and practices of global environmentalism from colonial and post-colonial histories.
8. In the late twentieth century media attention over 2000 'boat people' arriving from Asia verged on hysteria. The 'Asianization' of Australia' was blown up with little concern given to caring for displaced persons or their resettlement. Rather, media fueled questions were raised about arrivals, provoking a backlash among Australians fearing competition for jobs and housing. This backlash was ignited by Hanson's 'Asianization of a Nation' political marketing propaganda in the1990s.
9. See also Greg Garrard's (2012) *Ecocriticism* wherein he discusses Jeremy Bentham's early definition of speciesism from the eighteenth century.

References

Alaimo, S. (2019). Wanting all the species to be: Extinction, environmental visions and intimate aesthetics. *Australian Feminist Studies, 34*(102), 348–412. https://doi.org/10.1080/086164649.2019.1698284

Armitt, L. (2007). Dark departures: Contemporary women's writing after the gothic. In B. A. Brabon & S. Genz (Eds.), *Post-feminist gothic: Critical interventions in contemporary culture* (pp. 16–29). Palgrave Macmillan.
Byron, G. (2012). Global gothic. In D. Putner, (Ed.), *A new companion to the gothic*. London: Blackwell Publishing Ltd.
Bullock, E. (2017). Something in the dark: The tale of Ruby Rose and the Tasmanian gothic. *Metro Magazine, 182*, 84–91.
Campbell, J., & Hamilton, S. (2017). They saw a thylacine. *Endangered: Three plays* (pp. 5–59). Currency Press.
Carleton, S. I., & Hassall, L. (2014). What is Australian gothic theatre? *Three Playwrights Enter the Conversation. NJ, 38*(1), 25–37. https://doi.org/10.1080/14452294.2014.11649570
Foley, M. (2020). Why is Australia a global leader in wildlife extinction? *Sydney Morning Herald*. https://www.smh.com.au/politics/federal/why-is-australia-a-global-leader-in-wildlife-extinctions-20200717-p55cyd.html
Forbes, J., & Spearitt, P. (2003). Rum corps to white-shoe brigade. *Griffith Review, 2*, 23–35.
Holloway, T. (2007). *Beyond the neck*. Playlab Press.
Hughes, L. (2021). Postnatural, post-wild, posthuman: Our troubled relationship with the blue marble, *Griffith Review: Remaking the Balance, 71*, 264–274. https://www.griffithreview.com/articles/postnatural-post-wild-posthuman/
Marquet, K. (2018). *The dead devils of cockle creek*. Brisbane: Playlab Press.
Robin, L. (2010). Battling the land and global anxiety: Science, environment and identity in settler Australia. PAN: *Philosophy, Activism, Nature, 7*, 3–9. https://core.ac.uk/reader/36963890
Varney, D. (2015). "Beauty tigress queen": Staging the thylacine in a theatre of species. *Journal of Association for Australian Literature, 15*(2), 1–13.
Ward, M., Simmonds, J., Reside, A., Watson, J., Rhodes, J., Possingham, H., Trezise, J., Fletcher, R., File, L., & Taylor, M. (2019). Lots of loss with little scrutiny: The attrition of habitat critical for threatened species in Australia. *Conservation Science and Practice, 1–13*,. https://doi.org/10.1111/csp2.117

CHAPTER 5

The Mythic Space

Abstract This chapter explores the intrinsic relationship between the people and the places mythologized in Australian landscape culture since colonization. The mythic identity lies within the psychological and geographical hardships encountered during settlement and colonization, and from a Climate Gothic perspective, the tension between both states enhances the obsessive quest for land ownership and the desire for a place to call home. The chapter discusses how the Australian mythic space is in reality, conjecture, an inauthentic representation of a collective psychology that has informed the perceived national identity. The following analysis suggests that the space, so strongly associated with popular images of the desert, the outback and the bush, is both an oversimplified *and* an over-exaggerated representation of a more complex and often corrupt contemporary reality. The chapter explores these concepts through the Climate Gothic lens of two plays, *The White Earth* (2007) by Shaun Charles and Andrew McGahan and *The Flood* (2012) by Jackie Smith.

Keywords Myth · Myth-scape · Regional Legacy

Mythic depictions of the actual and the cultural Australian landscape have been the canvas for a persistent national obsession about who we are and our place in the world. Derived from a preternatural continent of

L. Hassall, *Theatres of Dust*,
https://doi.org/10.1007/978-981-16-6159-4_5

dialectic oppositions, Australians have enacted idealized and mythologized essentially masculine versions of the transition from convict roots to settler nation. Rather than projecting insight into the historical activities that affected contemporary environments, the landscape became an adversarial symbol of our imaginary relationship to it and therefore lacked empirical representation of its genuine existence. In the twenty-first century, concepts of the Australian space have changed; space has become a place to adopt, refine and question long held and very popular myths about family, identity and nation, which have affected the larger climate-affected world. The following discussion suggests that the mythic mapping of the landscape, its adversarial representation and its effect on family and identity was over simplified. Further, the discussion suggests that a complex contemporary reading of the tension between past and present sits between internal fantasy and the external reality of the present experience. The nostalgic desire for the past's imaginings of place, to set down roots and start a family in and of country are redundant, superseded by speculations about a climate-affected future where mythic ideals traditionally projected to the wider world are challenged or corrupted in deteriorating physical and psychological environments.

When discussing the Australian myth as stimulus for the Climate Gothic, the heroic representation of people and the mythic depiction of these people in forging the white nation commands some attention. The glorifying of people, place and land within historical depictions enacted a mythic space well into the late twentieth century. Australian myths about identity and nation developed as a way of shaping a more sanitized alternative to white history (Hodge and Mishra, 1991). Our obsession with developing a mythic narrative about this country's beginnings align with Robin's (2010) theories concerning identity and non-identity. Robin posits that 'unlike Indigenous Australians, non-Indigenous Australians are "non-identified" mak[ing] … a person invisible … fluid … or ambiguous: the state of mind that imposes itself upon, rather than lives through the landscape' (p. 5). The defining characteristics of the white Australian space have been of an ancient land, chimerical in its conflicting representations. Vast isolating distances hypnotically call to us. Its untouched beaches, sweeping plains and immense rich grazing land fuel our obsession for land. Its red desert heat is part of us. The glorifying of people, place and land within these depictions enacted a mythic space well into the late twentieth century. Chartered almost exclusively in terms of human values and experience, the seductive mythic landscape (myth-scape) interprets a resource for either human labour, appropriation and/or profit (Jordan, 2014). Our historical relationship with the landscape, popularly mythologized in art, literature and drama,[1] is now analysed from the perspective of the contemporary climate crisis in which we are living (Urban, 2007). The critique of historical representations of the much-loved myth-scape suggests a fictional space, construed through white socio-historical memory that eulogizes the heat, the distance, and the land to its own advantage (Hassall, 2020). Contemporary Climate Gothic analysis of our relationship with the landscape, of how it has defined us, suggests a more sceptical view of history and its effects on the contemporary space. It problematizes mythic depictions as misrepresentations perpetuated by mythic *unreason*, and the national obsession with defining a unique sense of identity to the wider world. Extending on previous Climate Gothic discussions, the preoccupation with mythologizing a heroic, settler past weaves through a contemporary investigation of colonizing processes deemed detrimental to the landscape that ironically cultivated the myths.

The ideological myth-scape of the past proposes a frontier country, an entirely (white) masculine terrain that promoted an almost pathological stereotype, in a place that in reality was almost too much to bear. Australians have perpetuated such popular narratives, indeed taken pride in tall tales of the 'Little Aussie Battler',[2] the (white) man on the land who was always trying to 'beat the bloody odds',[3] persistently at war with the cockies,[4] the law or the land that saw the heroic battler often defeated, over-awed and overwhelmed by the country itself. Man, and it is always man,[5] combats geographic and physiological hardship within the myth-scape. His battle is a cause of celebration, a source of national pride. The Aussie male's ability to try and beat anything that the 'rich bastards' or 'the bitch of a country' throws at them, including dust, dirt, drought, weather, mongrel cattle and blunt axes, bludgers[6] and brown snakes and crocodiles has encouraged a popular stereotype to the rest of the world. The character types that emerged supported the mythic representation of place and people, laconic rebellious men, stalwart and strong, who fought for the pioneering life even as they fought against the land. While society *en masse* could not make a significant dent on the ancient land itself, Australian folkloric history has obsessed over the heroic feats of settler activity in a hostile country, and as Gibson (1992) remarks:

> … the heroes' persistence in the legendary setting persuades us that the land is habitable but only by a very special breed of people [white men]; in mythic terms, therefore the nation is feasible, but only as a collection of extraordinary [male] individuals. (p. 73)

These working-class heroes are conjured in the collective imagination. Their dogged experiences of class conflict (Shearers Strikes[7]), failed revolution (The Eureka Stockade[8]) or their despotic but unflagging loyalty to a white settler nation (The White Australia Policy[9]), firmly embed the mythic legacy associated with the masculinized nation. Experiences of forging the Australian frontier have developed into a vocabulary of nostalgia … informing a narrative immortalized in the tall tales of the past, yet inaccurate in historical authenticity. The bush and outback landscapes and the pastoral depictions of the early Heidelberg School of painters of the nineteenth century[10] mythologize haunting, light or mist infused visons of the country. Our early playwrights of nineteenth and twentieth centuries[11] are fascinated with describing regional Australians through their relationship with and to the land. The reality of colonization and the activities of nation-building became secondary to narrating our imagined place in the country's vast distance and isolating spaces. For two centuries, Australian artists overall made little attempt to represent the effects of sheep and cattle farming on the land nor did they attempt to depict the impacts of fencing and farming land on First Nations People's relationship with country. Indeed, the invasion of cattle and sheep became part of the occupation which disposed of Aboriginal peoples of their lands, and which contribute to the escalation of soil salination, soil erosion and habitat destruction.

The myth-scape in which the Australian identity was established represented a retreat or escape from what playwright scholar Stephen Carleton (in Hassall, 2014) describes as 'all things urbanized, an absence from culture and reality' (p. 130). It is the city perspective that extolled our heroes and as a result, made the mythic Bush or Outback possible. The mythic imaginings of the past are now fleeting fragmentary images, nostalgic invention, offering a hint of what has passed, an enactment of a tall story from a misremembered childhood—a rendering, or a translation of events that expose our national fantasies about what is 'out there' in all that empty space. The fantasy illuminates how, as a white developing nation, Australians defined themselves. Within the peculiarities of the landscape, they performed the Gothic paradox between loss and belonging, and the white Australian identity is woven through the tension between myth and history, e.g., white Australia defined itself as a colony that dedicated itself to establishing its uniqueness, despite established ties to the mother country and its brutal colonizing structures including terra nullius. As a result, the frameworks of settler society wherein the myths were born, suggest that suffering in surviving and controlling the landscape was a colonizing method of establishing Eurocentric rights to land ownership. The following discussion explores what is omitted from the mythic chronicles, essentially how land organization, control and practices in the white nation space have affected the ecological well-being of the Australian nation[12] and in turn, how the effects of these practices have changed the way the mythic character is represented in the landscape.

In a time of climate change, species extinction, rampant infection, political lunacy, economic crisis, religious extremism and terrorism, an analysis of the myth-scape can unearth the beginnings of volatile contemporary issues. Spatial reimaginings interpreted through Gothicized lenses argue notions about mythic place, space and race. Dramatic climate-scape representation and analysis offer a space to adopt, refine and question popular concepts of self, family, nation and place attachment. It is evident that Australians have a very strong sense of place attachment. Place attachment as described by Harris (2008) is indicative of experiences of nostalgia, homesickness, patriotism and nationalism. The symptoms described embrace the role the myth-scape plays in the physic investment of the Australian identity. The following discussion draws on these concepts within the duality of the Climate Gothic space. This space *informs* the aesthetics of new climate-scapes and further *questions* how spatio-climate experiences contribute to constructing meaning about our changing identity within it. The plays discussed in this chapter situate the myth-scape in past memories of an external panorama, memories of a time marked by the borders and boundaries put in place in the Australian quest for domestication and ownership of land and in the setting down of family roots. The first play discussed is *The White Earth* (2009) by Shaun Charles and Andrew McGahan.[13] The play inverts the beloved myth-scape by bringing to the fore questionable beliefs about land as a divisible commodity. Themes of white masculine obsessions over land value, ownership and legacy are paramount. The psychological space is a contested racial space and a space that excludes women (aligned with traditional depictions) in the masculinized place-making processes. The climate-scape of the play is therefore, both geographic and psychological, a place where dreams of home and taking ownership of the land reign supreme. However it is also a space wherein the construction of *home*, in true Gothic mode, houses 'the dark dreams and desires of a male white nation [wherein] the monsters ghosts and madmen reign' (Charles & McGahan, 2009, p. 5).

As discussed in earlier chapters, it is difficult to separate Australian Gothic and indeed Climate Gothic from landscape issues that embed colonizing practices that have been detrimental to the First Nations Peoples of this country. *The White Earth* challenges our white obsession with setting down roots and making a 'home' in the landscape. The play inverts the 'Little Aussie Battler' mentality of finally 'beating the bloody odds' as themes of place are investigated through the Native Title debate surrounding Indigenous land rights claims.[14] The anticipated threat that the historic Mabo land rights decision has on the lead character, JOHN McIVOR's right to the contested land, foreshadows his final escalation into madness. Extreme drought and unpredictable fire events expose the now parched and burning myths concerning land inheritance, and the play blatantly expresses disenchantment with idealized systems and functions associated with colonial claims of terra nullius.

The White Earth situates a drought and fire affected landscape as a place for positive change in race and land relationships. It is a classic white Australian ghost story; the haunting psychological atmosphere reverberates with the hysterical racial anxieties of a 1990s white Australia in conflict over land rights. The story begins in 1845 with the raising of Kuran House by the settler WHITE family (pun definitely intended), who followed explorer Alan Cunningham's trail into Queensland's Darling Downs, a farming region of rolling grasslands and rich soil. Other settlers followed, herding cattle and sheep before them. In this fertile place, the WHITE family reared a great homestead; Kuran House, an Australian version of Gothic mausoleum-esque architecture and nostalgically reminiscent of the great mansions of Mother England, is a mansion of slate and sandstone that housed generations of WHITEs to follow. In its glory days, the Station[15] stretched twenty miles north to the south and twenty-five miles west out to the plains. The house is now a collapsing ruin, devoured by the land on which it stands. Kuran is the overarching dramatic symbol, dramaturgically positioning debates about terra nullius within a distinctly white Australian spatial experience. There is no doubt that Kuran is an antagonist in the piece, a reminiscent ghostly symbol of settler glory, a built environment that layers the colonizing processes of the past into the politically fueled racial agenda of the present. Deeds of ownership, while paying testament to mythic representations that applaud 'beating the bastards', also symbolize the fear of losing ownership rights. It is this paradox which drives the tension and dramatic conflict.

At the time of the historic Mabo decision, JOHN McIVOR, a patriarchal, half-demented recluse retains ownership of the deteriorating property. JOHN grew up on the Station, not as an owner but as an employee of the WHITE dynasty of Pure Merino sheep barons. His father DANIEL believed JOHN would eventually marry the granddaughter of settler family, ELIZABETH WHITE, and thus change the McIVOR fortune. However, when her father and her despotic, alcoholic brother die, ELIZABETH sells Kuran, leaving the McIVORs destitute and without a home. The 'Little Aussie Battler' DANIEL McIVOR is now an exile from the land he covets. His fury at what he believes is an unjust expulsion drives him to alcoholism and eventual death. The forced removal further infects DANIEL's son JOHN, who begins a never-ending quest to gain back the land he believes is his legacy. Throughout the years of his exile into his years of maturity, the idea of Kuran Station possesses JOHN. His identity is inexorably linked to the Station. After the death of his father, he visits the land on which he youthfully roamed. A demon awakens in JOHN and his quest to take possession of the house begins in earnest. From a Gothic perspective, the house is a symbol, a frightening tomb to white colonization processes. JOHN McIVOR represents the colonizer crippled by his own greed. Another death sets the current events in motion. JOHN's estranged relative dies a fiery farming accident. With nowhere else to go, WILLIAM (the surviving nephew) and his mentally declining mother, arrive to live at Kuran. WILLIAM's first glimpse of the house embraces the Australian Gothic perspective:

> NARRATOR: Then at last, Kuran House was revealed through the trees, and William felt a deep chill of premonition run through him. It seemed that in that first moment, that it was not so much a home he saw, as it was a great broken shipwreck, thrown up on the rocks from the dry ocean of the plains below (p. 11)

The shipwreck, both symbol of convict transportation, subsequent colonization and metaphor for (post)colonial guilt, frames the haunted atmospheric presence of the contested racial space. The rage JOHN feels about his past exile from the Station has festered and he cannot see that he is perpetrating the same crimes against Indigenous titleholders who wish to access a sacred waterhole on the property. The waterhole possesses cultural and pre-invasion significance for them and is subject of JOHN's fears about the passing of the Mabo decision. It is also the site of an Aboriginal massacre, perpetrated by DANIEL McIVOR and witnessed by a young JOHN. It is the place where JOHN sees the BURNING MAN for the first time. The fiery vision haunts him throughout his life and is both a demonic accuser of crimes committed on sacred land and a prophetic vision of what we learn will be his own demise. The McIVOR's curse of exile, and later incest and madness, describes a mythic ***unreason***. The play offers an inversion of the 'beating the odds' myth and in doing so, suggests the mythic Australian masculinized narrative is in the final stages of decay. While the half-mad reclusive JOHN wants to pass on not only the Station, but also his almost supremacist zeal, WILLIAM, his recently arrived nephew, finds the rhetoric confusing:

JOHN: But let me ask you a question. This property, it's mine, isn't it? I own it. If I don't want anyone knowing about the waterhole [a sacred Indigenous site], for example, then that's my business. Right?

WILLIAM: Right.

JOHN: Well there are things happening in this country that will take that choice away from me if we don't put a stop to it.

WILLIAM: What things?

JOHN: Government things. Same as always. Interfering. That's why I started. The Independence League.

[John produces a newsletter…]

Here, this is my real work. It's this month's newsletter. The front page. Go on. Read it.

WILLIAM: *[Slowly.]* 'We reject the United Nations and any other body that seeks to limit Australian sov … sov …'

JOHN: Sovereignty. Good, keep going.

WILLIAM: 'We reject excessive immigration.'

JOHN: You bet we do.

WILLIAM: 'We reject the alienation of Australian soil to minorities.'

JOHN: These are grim days, Will. The Mabo judgement. This new Native Title legislation that they're bringing in. They're disasters.

WILLIAM: 'We believe in an Australia run by all Australians for the benefit of all Australians.'

JOHN: I've organized a rally for the members, right here on the station. There's going to be hundreds of people. They'll come from all over. Country folk. City folk too. Good honest citizens.

WILLIAM: 'We believe in one flag.'

JOHN: Yes.

WILLIAM: 'In one people.'
JOHN: Yes.
WILLIAM: 'One Nation.' (pp. 33–34)

The Australian League of Independence has 'quasi-seceded from Australia' (Starford, 2020, n.p.)—a country that, it believes, is undermined by the liberal advances in land rights for Indigenous Australian people/s and styled on Pauline Hanson's One Nation party which promotes a white 'mono-cultural and mono-social nation' (Starford, 2020, n.p.). In a united white One Nation, identity is located within empirical processes enforced onto an Indigenous culture-scape.[16] McIVOR's obsession and possession stem from settler anxieties grown in disputed racial spaces. Escalating drought and fire events are represented in the fierce and uncompromising racial zeal represented by the League. The threat of fire is deeply embedded in the Australian consciousness, and the flaming apparitions that appear herald the blazing resurrections of the dark, demonized victims of history that fuel the League's racial anxieties over Native Title claims. The oppressive heat, the drought and the constant threat of fire, the vicious wind and the visions of BURNING MAN further promote the Climate Gothic reading. The overwhelming ferocity of these elements suggests that the homestead and all it entails is a delusion, the least significant aspect in the climate-scape, a gaunt and skeletal shell encompassing a lifeless space peopled by ghosts and dead dreams of land.

After WILLIAM takes up residence, the various visitors and visitations, ghosts and sightings of those long dead, pass through the house, each ghostly spectre belying the mythic fantasies that JOHN strives to pass on to his lonely and isolated nephew. WILLIAM spends his time roaming through the empty house, exploring tattered and moldering trophies of the past splendors of the homestead. These trophies, the rotting house, the drought-stricken land on which it stands, the extreme weather and the terrifying BURNING MAN are to be WILLIAM's legacy.

NARRATOR: And so it was that the westerly winds began to blow. Hot and dry and fierce, the scalding breath of summer come early and hard, and this third year of drought in a row. Day after day the wind beats about the House. Trees bent and shuddered, and out in the hills, the last blades of grass withered and died. In the mountains, fires began to burn. For William, the wind's arrival marked a point where the world started to slip away. A dislocation was creeping over him – an illness,

> a dizziness, a constant ache in his ear, and there was an evil smell that clung to him, more sweet and sickening every day. At night, he dreamed of a man on fire. A man who only stood silent as he burned, watching. (p. 63)

Regional environments such as the one depicted in *The White Earth* are successful hunting grounds for both Gothic and Climate Gothic analysis. *The Flood* by Jackie Smith (2012) is thematically similar to *The White Earth*, moving beyond a purely thematic investigation of an Australian environmental legacy by viewing the mythic ideology through another fractured farming family lens. The Climate Gothic imagination in both plays investigates an Australian obsession with land ownership and the repercussions of such on family legacies. It suggests that certain depraved deeds committed within the fenced climate-scapes have significant repercussions on family legacy. Importantly, the psychological legacy of evil deeds perpetrated within families in isolated farming environments is that which affects ecological custodianship. Similar to *The White Earth* is the positioning of the homestead, a symbol of the disappearing pastoral dream of a sheep farming empire long past. *The Flood* is set in a deteriorating farmstead in the unique Gothic climate-scape of rural Tasmania. The once grand family home is now going to rack and ruin. Ensconced within the filthy walls, JANET and her two grown daughters CATHERINE and DOROTHY reunite and watch the river that courses through the front of their property, flood. The weather escalates the atmospheric *dis*-ease embedded in the relationship between the mother and her daughters. The three women are the last of the once middle-class farming family, and while JANET has remained on the farmstead, her daughters have struggled to escape the past and the oppressive pastoral family traditions.

The incessant rain and the rising floodwaters set the backdrop for the escalating tension simmering between the three women. The climate-scape is ominous, across the island state the weather is erratic, flooding in the East while bushfires rage in the West. The forced confinement of the women within the filthy homestead enhances the haunting atmosphere as the weather outside rages and the once 'safe-space' of the home reveals its dark doppelgänger. The Climate Gothic reading sits within the tension between what is familiar and simultaneously unfamiliar in the geographic surroundings that frame the Femi-centric family reunion. There is no reprieve from the weather, or each other, as roads are cut and communication with the outside world breaks down. While the farmstead is situated only 40 kms from the nearest town, the women might as well be buried as deep in the Tasmanian wilderness as the characters of *They Saw a Thylacine* or *The Dead Devils of Cockle Creek*.[17] The atmosphere is uneasy. The rain is relentless. The farmstead that was once a testament to the sheep farming empire of Australia is now a shadow of its former glory. Any remnants of the farmstead as a healthy symbol of the myth-scape tradition are gone:

> CATHERINE: It's raining.
> JANET: It never rains.
> CATHERINE: It does now. The kangaroos are drinking from the puddles.
> JANET: Vermin. Should be shot. They eat the sheep's food.
> CATHERINE: I didn't see any sheep.
> JANET: All sold up now. No sheep. (Smith, 2012, p. 4)

The women, thrust into isolation must confront their family demons as secrets thought drowned in the murky depths of the flooding river, resurface. JANET sits in a nest of tatty blankets and cushions, mumbling away as she reminisces about her days as a young wife and mother, and cutting fresh images of pretty, clean, white people from the paper detritus. CATHERINE the estranged daughter returns from civilized exile in London after hearing of her mother's deteriorating mental condition. She is disgusted to notice that the once magnificent home of her childhood is littered with photographs, magazines, catalogues, old cups, books, dirty crockery and food scraps. From the outset, we realize JANET's ghosts are closely linked to her obsession with, cutting, collecting and collating clean white, yet sub-textually erotic images from the mounds of paper litter. JANET, like JOHN McIVOR, lives in a state of limbo with the ghosts of her past. Trapped in deeply buried family secrets concerning her husband's incestuous ravaging of DOROTHY, she lives her own lies about how she dealt with his pedophilic infidelity. While the Gothic is renowned for its exploration of 'sexuality' that is harrowing in its aberrant nature and associations with perversions of power,[18] the two plays discussed in this chapter corrupt this aberrance further through the sexual exploitation of children, and it is the ghosts of these memories that haunt JANET and JOHN in their separate climate-scapes:

> JOHN: I should have killed you that night.
> DUDLEY: I'm so sorry…
> JOHN: I should have killed you that night.
> DUDLEY: Please, I didn't mean to …
> JOHN: I should have sent you away.

> DUDLEY: Please where can I go, I can't go anywhere, I'm so lonely John, I can't be all on my own…
> JOHN: God help me you can sleep in the shed … (*The White Earth*, p. 51)

Even when JOHN discovers the terrible secret, he is incapable of action, therefore acknowledging the abuse will continue. Incest goes against everything that our foundational myths build upon, e.g., solid family values and protecting innocent children. Questioning these values suggests a vital element of Australian mythology is redundant, ruptured in cycles of family abuse. DUDLEY is JOHN's closest and trusted friend, an 'uncle' to his daughter RUTH. Uncle DUDLEY is a shambling, filthy, half-crippled WWII veteran. In essence, he represents the antithesis of the ANZAC legacy, the anti-hero. JOHN proves incapable of running the pedophile off the land and we learn that DUDLEY is in fact, co-owner of the property. If DUDLEY is exiled, JOHN fears he will lose controlling interest in the property. In a cruel, twist of fate, the child RUTH is the one exiled, sent away to boarding school in the city. The consequence of RUTH's abandonment is in the rupturing of the father/mother/daughter relationship ***and*** in the breaking down of the family's productive relationship with the land. Thematic similarities between the two plays are evident. Both explore how the breakdown of the family unit affects the well-being of the land and its productivity in the sheep farming culture. In *The Flood*, we learn CATHERINE is sent to boarding school. Exiled to protect her from her father BRIAN, she is unaware that he has been sexually interfering with DOROTHY. At the time of *The Flood*, BRIAN, is 'presumed' to be dead, the victim of a similar weather event in 1973. In both plays, we learn that sexualized horror happened at the peak of both homesteads' functionality. In both plays, geographic hardship and social isolation promote eroticized dangers. In both plays, there is enough distance from civilized society to ensure there is no intrusion of acceptable standards of morality concerning forced sexual activity with minors. According to Foucault (1978), incest is a threat that generates an implicit alliance between family members. Sexuality within the family supports a generational legacy of family secrets. Foucault also suggests that the rise of incest discourse aligns with the development of capitalism and the transmission of wealth from one generation to the next. Ironically, the Gothic paradox sits in both JANET's and McIVOR's failure to act. Their inability to protect and nurture their children is the root of the family's, and consequently the land's, deterioration. The incestuous acts ensure that there will be no moral legacy of succession and male heirs, if any at all, will not preserve the man-on-the land mythic tradition. In both plays, the homesteads symbolize status and power, the male domain, the Gothic castle wherein the women are mere physical and psychological prisoners trapped within the confines of male plundered myth-scape. DOROTHY explicates:

DOT: This place has a way of trapping you, whether you're here or not.
CATHERINE: I've never thought about here as being trapped. I liked it when I was little. The horses and dogs and stuff.
DOT: Hello Fucking Anne of Green Gables. You hated this place, you were always terrified of it ... when you came back for holidays you hated the dark ... (Smith, 2012, p. 39)

Fear of darkness becomes a motif in the present circumstance. As the sisters reminisce, darkness falls and the rain suddenly intensifies, as does the atmospheric layering of hauntings in the climate-scape:

JANET: There are funny things out here. I've never understood half of what goes on. But there are things, spirits, don't you think?
CATHERINE: I suppose.
JANET: You're never alone, even when there is not a soul for miles. You never feel alone out here, there's always something, watching. Sometimes, I'm out there by the river, and I feel like someone's just behind me, and I turn but there's no-one there.
CATHERINE: It gets lonely.
JANET: There's other things, things you can't explain. You can watch, on the surface of the river, how it swirls sometimes, but there's nothing, nothing there to make it swirl. Not a stick, not a branch, a tree,
CATHERINE: Maybe underneath the water?
JANET: No, there's nothing. Or you just see a lump on the branch. Just a plastic bag, or is it something yellow, a dead thing all caught up.
CATHERINE: Mum!
JANET: Listen – no sound. Not a buzz or bird, not a breath of wind. Only the river and the rain and the water. (Smith, 2012, p. 15)

The weather, taking on distinctly human characteristics, is the overarching energy force. The silence, the absence of signs of life, enhance the Climate Gothic atmosphere. The *absence of life* opens the doorway to otherworldly visitors. The personification of the environment haunts the climate-scape and it taps at the window, crashes at the roof and knocks at the door:

CATHERINE: Mum said there's a Kraken out there.

> DOROTHY: What the fuck's a Kraken?
> CATHERINE: A sea monster.
> JANET: If you poke the mud up with a stick, it rises to the top.
> CATHERINE: What does?
> JANET: Whatever is at the bottom.
> ...
> Ssh. I'll get the gun. (Smith, 2012, p. 64)

The Kraken, perhaps symbolic of the monster BRIAN, needs to remain buried in the flood-affected climate-scape. JANET later reveals she has killed BRIAN in 1973, shot him through the stomach, dragged him behind the tractor and pushed him into the fury of the rising river. She has let the land take hold of him, let the river remove all trace of him from the morally corrupt landscape in which he abused her daughter. However, in an ironic twist, we learn that BRIAN has actually fled the farm, never punished or taking responsibility for his sexually corrupt acts. JANET, imprisoned in her murderous lie, reveals her desire to *believe* she acted in the best interest of her daughter. For JANET, like McIVOR, there is no redemption. She will remain trapped on the land that holds the secrets of the sexual and psychological violence that determined their present fractured relationship. Herein lies the Gothic paradox; BRIAN, the monster, is not dead, but has fled the family and land he ruined, to live in the city, father a new family, and re-enact the contradiction between father and perpetrator. BRAIN is the symbol of colonization, and the raping of his daughter becomes a metaphor for the colonizer's raping of the landscape. Past and present atrocities collide in the corrupt act. The Gothic quilting of debased experience disrupts the myth-scape tradition.

The acts of incest described enhance the Climate Gothic atmospheres in both plays as they occur within a fire and flood affected melting pot of other sins. The hallowed 'man-on-the-land' heroic legacy ruptures, resulting in the shattering of the myth. Children become victims. Mothers become estranged. Families break. The land becomes infertile. The women become the passive receptacles of the male myth-making ideals in the morally corrupt landscapes of their own history. As dominant myth-scape settler symbols unravel, patriarchal power fantasies of sexual dominance become horrific reality. Both plays embrace the genetic features of horror and fantasy—fantasy cloaked in the questionable myths narrated in psycho-historical representations of colonization. While memories of the glory days of owning significant pieces of the country remain, both farmsteads rot around the characters, adrift in drought and flood and dark

Romantic legacies of incest and sexual corruption. The corrupted climate-scapes repudiate any ongoing desire for the fertile landscapes of the past and white hopes for what once was:

> JOHN: … This place is a piece of country. It has trees and scrub and creeks and gullies. It's three dimensional. It has its own life and its own history. It breathes and even talks.
> WILLIAM: It talks?
> JOHN: To the right person.
> WILLIAM: I don't hear anything …
> JOHN: Give it time, and you're not seeing it at its best. This drought … (p. 21).

The Climate Gothic readings that the bush and outback land-scapes provide in the discussed plays, posit new investigations of the classic Australian the myth-scape and the investigations further suggest that Australian playwrights are continuing the love affair with the land, though from more critical perspectives.[19]

Notes

1. For a comprehensive analysis of this concept in relation specifically to theatre, refer to Joanne Tomkins 2008 study of Australian drama, *Unsettling Space: contestations in contemporary Australian Theatre*. Tomkins' analysis refutes accepted spatial representations and constructions of the Australian historical space and consequent spatial identities.
2. A colloquialism that explores a self-deprecating, blue-collar character who is consumed with labor in *and* against the landscape.
3. See my article *Contemporary Performance and Climate Change* (*Australasian Drama Studies ADS*, 76, 2020) which discusses at length a white mythology that has developed from historical imaginings which embrace a national tradition wherein Australian-ness is directly connected to harsh psychological and physical landscapes.
4. Colloquialism for rich farmers who own thousands of acres of land for the purpose of grazing vast numbers of sheep or cattle.
5. When one refers to traditional representations of women in Australian fiction and drama, we see an isolated stalwart keeping the home fires burning, courageously battling wandering men, dogs or snakes while their men are off droving or to war or, a set of hard-bitten, nagging characters, who bemoan the weather, their lot in life and who fix most obstacles with a good strong cup or tea or a full-blown fight if need be. See John

McCallum's *Belonging: Australian Playwriting in the 20*[th] *Century* for a lengthier and interesting discussion on female representation.

6. Bludger is a colloquial term for a lazy bloke (male) who shirks hard work at every opportunity.
7. The Shearers Strike, which crippled the Australian economy in the early 1890s, is largely represented from a blue-collar working-class perspective by playwright Errol O'Neil in *On the Whipping Side. It* chronicles the drama and politics of the 1891 Queensland Shearer's Strike and challenges audiences to consider issues of workers' rights, fair representation, and the employee-employer relationship. Squatters, shearers, unionists, journalists, business owners, politicians, their families—all are represented in this sweeping epic of a definitive moment in Australian history.
8. The Eureka Stockade rebellion was caused by a disagreement over what gold miners felt were unfair laws and policing of their work by government. Police invaded the mines to enforce the licensing laws in late November 1854. The miners refused to cooperate, burned their licenses, and stoned police. The Eureka Stockade became a legend, not only because it was the birth of Australian Democracy, but because of the recognition of the courage, and determination of the diggers and their willingness to defend their rights.
9. The White Australia policy, formally Immigration Restriction Act of 1901, in Australian history, was a fundamental legislation of the new Commonwealth of Australia that effectively stopped all non-European immigration into the country and that contributed to the development of a racially insulated white society. It continued until 1973 when the Whitlam Labor government definitively renounced the White Australia policy and, in its place, established a policy of multi-culturalism in a nation that is now home to migrants from nearly 200 different countries.
10. The Heidelberg School was an Australian Art movement of the nineteenth century and is often described as Australian impressionism. The movement emerged in response to strong nationalist sentiment associated with federation and attempted to define Australian-ness through the landscapes depicted. Painters such as Arthur Streeton, Frederick McCubbin and Tom Roberts are considered key figures of the movement and they sought to capture Australian life through their depictions of the country, the bush, the sunlight that typified the nation and supported the growing national perceptions of identity.
11. See David Burn's *The Bushrangers* 1829, George F. Price Darrel's *The Sunny South* 1833, Oriel Gray's *Lawson* 1943, Douglas Stewart's *Ned Kelly* written for radio in 1944 and realized on the stage in 1956. These plays provide examples of the symbolic and mythic landscapes imagined associated with a preoccupation in defining a national identity. It should be noted that Oriel Gray's award-winning play *The Torrents* 1954 explored

feminism and saving the environment and was not received well by a largely conservative Australian audience.

12. See Gibson's (1992) provocative and now considered a definitive modern analysis of Australia's films in *South of the west: Postcolonialism and the narrative construction of Australia*, which examines an array of Australian objects and attitudes encountered in a southern Australian locale. The text interweaves a discussion to form an essay on the realignment of space, time, and meaning in contemporary Western societies. Gibson demonstrates how these different systems of representation construct 'Australia'.
13. The play is based on Andrew McGahan's (1966–2019) Miles Franklin Award winning novel of the same name published by Allen and Unwin in 2004. The play premiered at Brisbane's La Boite Theatre in February–March (2009) and was directed by co-author of the adaptation, Shaun Charles.
14. This was a time of enormous social and political change in Australian contemporary history whereby the Labor Government led by Prime Minister Paul Keating fought to make a law that recognized the unique and ancestral ties First Nations Peoples and tribes have to the land. This historic decision made on 3rd June 1992, saw the High Court of Australia decide that terra nullius should not have been applied to Australia. Steeped in national controversy, the decision recognized that Aboriginal and Torres Strait Islander peoples have right to the land—rights that existed before the British arrived and rights that can still exist today.
15. Station refers to house, external buildings and machinery, fenced land, and animals on that land.
16. Culture-scape is identified by all or some of the geography and by place, architecture, environment and social groupings; see Una Chaudhuri's (2002) discussion in E. Fuchs & U. Chaudhuri (Eds.). *Land/scape/theatre.*
17. See Chapter 4.
18. See Ruth Perry's (1998) excellent analysis of incest in Gothic literature in *Incest as the Meaning of the Gothic Novel.* The article draws on Foucault's theories of sexuality to analyze the importance of incest to a historical reading of the Gothic novel.
19. See McCallum's (2009) discussion on Australian playwrights and plays that have attempted to navigate the country through distinctions between city and bush psychologies of place and space.

References

Carleton, S. in Hassall (2014). What is Australian gothic theatre? Three playwrights enter the conversation. *NJ, 38*(1), 25–37, https://doi.org/10.1080/14452294.2014.11649570

Charles, S., & McGahan, A. (2009). *The white earth*. Currency Press.

Foucault, M. (1978). *The history of sexuality: Volume 1 An introduction* (R. Hurley, Trans.). Random House.

Gibson, R. (1992). *South of the west: Postcolonialism and the narrative construction of Australia*. Indiana University Press.

Harris, D. (2008). Self and landscape. In R. Z. De Lue & J. Elkins (Eds.), *Landscape theory* (pp. 187–194). Routledge.

Hassall, L. (2020). Contemporary performance and climate change: Redefining the Australian landscape narrative. Australasian Drama Studies, 76, April, pp. 270–295.

Hodge, R. I. V., & Mishra, V. (1991). Dark side of the dream: Australian literature and the postcolonial mind. Allen & Unwin.

Jordan, D. (2014). *Climate change narratives in Australian fiction*. Lambert Academic Publishing.

McCallum, J. (2009). *Belonging: Australian playwriting in the 20th century*. Currency Press Pty Ltd.

Robin, L. (2010). Battling the land and global anxiety: Science, environment and identity in settler Australia. PAN: Philosophy, Activism, Nature, 7, 3–9. https://core.ac.uk/reader/36963890

Smith, J. (2012). *The flood*. Tasmania: Australian SCript Centre, AustralianPlays.org

Starford, R. (2020). On 'The White Earth' by Andrew McGahan. Griffith Review, 70: Generosities of Spirit. https://www.griffithreview.com/24754/

Urban, K. (2007). Ghosts from an imperfect place in Phillip Ridley's Nostalgia. *Modern Drama, 50*(3), 325–345.

CHAPTER 6

Romantic Legacies: Ecocriticism in Australian Performance

Abstract This chapter draws on eco-critical discourse to frame the discussion on Stephen Carleton's award-winning play, *The Turquoise Elephant* (2016). Eco-critical analysis in contemporary dramatic discourse and of theatrical representation has grown from the literary legacy of Romanticism, specifically in fields of landscape aesthetics and writing about nature and the natural world. Initially providing an overview of ecocriticism and its corresponding relationship with Romantic legacies, the chapter acknowledges that the increasing visibility of ecocriticism in drama and theatre studies creates a successful framework for Climate Gothic analysis as it engages with the interrelationship between theatre environmentalism and climate change. Carleton's play draws on and inverts legacies of eco-critical Romantic traditions through its Gothic paradoxes of light and shade and good and evil. It is within this paradox of inversion wherein the Climate Gothic reading laments the loss of the nation's connection to non-human nature. *The Turquoise Elephant* question the values that we attach to the environmental crisis in this country.

Keywords Ecocriticism · Cosmopolitan Gothic · Romantic Legacies

Throughout history, artists from a variety of disciplines have been at the forefront of recording and reflecting on the unique natural world within

L. Hassall, *Theatres of Dust*,
https://doi.org/10.1007/978-981-16-6159-4_6

which Australian society exists and upon which we have imprinted our human stamp. Climate change requires Australian dramatists to question traditional depictions of writing about nature and the environment, and the dramatic shift in perception often draws upon literary elements popularized in the Romantic period. The Romantic period was the popular European literary form at the time of European convict transportation to Australia, and Gothic literature was a development of Romanticism. The Romantic writers of the past explored nostalgic imaginings of non-human nature. The incorporation of these elements into contemporary drama implies an eco-critical framework, which is useful for analysing socio-cultural and political-economic themes in drama studies. Ecocriticism is a legacy grown from Romanticism's literary fields of landscape aesthetics and writing about nature and the natural world.[1] Glotfelty's (2006) definition of ecocriticism is perhaps the most widely recognized and simply states that it investigates the association between literary works, i.e., literature and/or drama and the environment. An eco-critical analysis of Climate Gothic expresses a collective *dis*-ease with rational arguments contradicting climate science and considers the emotional anxiety and uncertainty associated with the warming of the planet. Therefore, Climate Gothic dramatic analysis sits on some level within the analytical category of ecocriticism (Trexlar & Johns-Purta, 2011; Jordan, 2014; Hassall, 2017, 2020).

The increasing visibility of a range of theatre-related activities that engage with the interrelationship between theatre environmentalism and climate change, suggests eco-critical dramatic analysis is a developing criticism responding to gaps in cultural climate knowledge[2] communication. The following discussion looks at how ecocriticism embeds Gothic paradoxes of light and shade and good and evil within Climate Gothic climate-scape(s) analysis. It is within this paradox of perception wherein *The Turquoise Elephant* (2016) by Stephen Carleton draws on and inverts literary legacies of Romanticism, as it laments the loss of the (post)modern individual's temporal and spiritual[3] connection to non-human nature. Disappearing Nature is a source of entertainment, a symbol of society's corruption and a sign of selfish human indifference to the other-than-human world. Further, the play defies post-enlightenment notions of rationality through its exorcism of moral questions associated with extreme climate change experiences. In true Gothic style, the play celebrates the irrational and the immoral. Projecting a darkly satirical exploration of human response to climate change, the play draws on fixed belief systems and familiar orientation points in an Australian cosmopolitan climate-scape. The cosmopolitan setting is especially disconcerting as it utilizes capital city, iconic orientation points as predominant climate-scape sites. Further, it hazards that the entire Australian population is in peril, one vanishing species among many. Carleton successfully draws on his Gothic expertise to build the Climate Gothic dramatic world.

The Turquoise Elephant is an exceptional Australian example of the way physical, psychological, perceptual, producible and imagined spatial ecologies are situating themselves in the eco-critical field. *The Turquoise Elephant* reassesses Australian values through its examination of the luxuries of an irresistible (post)modernity in time of climate change. An eco-critiquing of the work includes a dynamic re-evaluation of privileged globalized systems as Carleton satirizes excessive over-consumption. It is a cautionary tale of excess, exhibiting what Heise (2008) discusses as an eco-cosmopolitan environmentalism. Within the world of the play, a series of absurdist, almost operatic nightmarish futuristic scenarios occur, specifically in the cultural discourse of two major Australian cosmopolitan world cities. By putting the terrifying effects of climate change in familiar Australian locations—the cities of Sydney and Melbourne—Carleton depicts an eerie global climate-scape, and makes two of Australia's renowned capital cities terrifying and, unfamiliar wastelands of our own making.

The eco-critical purview within the *The Turquoise Elephant* suggests climate change enacts a specific political, interactive role within the socio-cultural milieu. Climate-affected Nature in this play is a tangible 'presence' and the catastrophe-affected cityscapes take on almost symbolic functions as non-human antagonists. Ironically, the climate-scape examines humankind's awe and even surprise in the face of their loss of unity with non-human nature, while simultaneously, the exploitation of financial and political opportunities resulting from the climate chaos is satirized. What is important about this play from a broader perspective is that it clearly embraces Australian Gothic principles that examine the dictates of history, while as a Climate Gothic representation, it considers the volatile effect of present and future politicized environmental factors on contemporary and near-future life. Inverting Romantic principles that viewed humans as pure and good in Nature's spiritual realm, Carleton undertakes the task of aligning the characters' moral and ethical outlooks with absurd socio-political views unsympathetic to environmentalism. He satirizes both the characters' and the society's lack of concern about the loss of non-human nature. There are deep levels of sub-textual complexity in the play's examination of the personal, social, political and active choices that we as individual Australians make when responding to climate change. Carleton's mastery in the form is evident, realized in the subtle, yet complex movement between Australian Gothic and Climate Gothic, wherein reflections of the colonial foundations of the country weave throughout the current perilous circumstance.

One of the great strengths of this work is its ability to satirize talking head events that advocate response to unrealistic demands on the planet:

> AUGUSTA: We need to get back to fossil fuels. Coal. Uranium. Now! Renewables are what got us into this mess. Solar panels are useless if we're about to plummet into an Ice Age. No sun. No power.
> BASRA: That's crazy.
> AUGUSTA: No dear. Perfectly sensible. (Carleton, 2016, p. 6)

The resource debate and its proposed solutions to the climate crisis serve as a powerful metaphor for the dissolution, doubt and/or self-satisfied indifference to the rapidly deteriorating climate-scape. A World Business Summit (WBS) is the backdrop to the escalating Australian climate crisis situated in the dramatic climate-scape. Set somewhere in the near future, we learn that we are beyond the tipping point, that rising sea levels are inundating Melbourne's sewerage pipelines. Its underground rail transport system is flooded and there is mass panic as Melbournites predict a potential cholera epidemic. Temperatures have escalated to the high 40s and low 50s degrees Celsius. AUGUSTA MACQUAIRE, descendant of Australian penal colony royalty,[4] will propose a Treaty at the WBS on climate-related issues condemning the breakdown of the Australian Green movement:

> AUGUSTA: We're challenging the science on climate change once and for all. Killing it dead
>
> [...]
>
> We've given the recyclers and wind farmers a quarter of a century to solve this. It's only gotten worse. The Green movement has failed. Worse than that. It's lied. We've found the evidence. (Carleton, 2016, p. 5)

The Treaty will propose a reintroduction of mining rights and advocate for extensive use of fossil fuels as a means of addressing the environmental crisis faced by the Australian population. The Governor General is proposing all coastal city communities relocate to higher ground. The global perspective is even worse. The last of the permafrost has disappeared in Greenland, the last of the snow is melting on Kilimanjaro, and in Cairo, the Sphinx is about to be submerged under rising floodwaters. Within this absurdly terrifying global climate-scape, an eclectic cast of egocentric and obnoxious characters engage in varying levels of eco-criminal activity. This collection of unlikable characters typifies everything the average working-class Australian despises. There are the surviving MACQUARIE women, nightmarish relics of a family who, for generations, reaped the rewards of 'riding on the sheep's back'.[5] Then there is JEFF, a big business property tycoon and con artist reminiscent of renowned Australian con man Christopher Skase.[6] He is plotting to acquire the MACQUARIE family fortune through nefarious means. VISI is a member of The Cultural Front for Environmental Preservation (CFFEP), a terrorist group disguised as artistic-environmental revolutionaries. Hired by the MACQUARIE family, she is disguised as the domestic help. BASRA MACQUARIE is an agoraphobe, who makes the best of her disorder by editing 'The Turquoise Elephant', an environmental activist blog she pens from the safety of her climate-enforced fortress home. These characters debate climate-related values and decisions that affect their immediate status quo and, in an abstract way, the continuing status quo of country. Their arguments represent the broader Australian divide between personal perspectives and the larger forces of culture, society, politics and the environment.

In its parodying of the populations' response to climate-affected Australian places and spaces, *The Turquoise Elephant* houses its thematic inquiry in a broader global investigation. Uniquely, the play satirizes personalised concerns that are localised within a broader globalised sense of place Heise's (2008):

AUGUSTA: Where's the maid?
BASRA: The Philippines.
AUGUSTA: Cecilia?
BASRA: Left on Sunday.
AUGUSTA: Not another one of those hysterical Catholic festivals?
BASRA: Her mother-
AUGUSTA: Some sighting of the Virgin Mary on a garbage mountain.

BASRA: Her family has just been wiped out by a category six typhoon.
AUGUSTA: Not again.
BASRA: They've lost everything.
AUGUSTA: When's she coming back? (Carleton, 2016, p. 2)

Pivotal to any eco-critical fictional reading of the climate change themes explored is the possibility of the realistic likelihood of the circumstances occurring (Picart & Greek, 2010). While Jordon (2014) suggests that 'realism often fails to address Anthropocinism' (p. 42), Carleton draws on, at least to some level, a degree of authenticity or realism to heighten the absurdist nature of the character's extreme behaviours. The climate-scape offers a new way of reflecting on the Australian psycho-political landscape. In doing so, it provides a unique theatrical examination of the ever-present problems of consumerist evils that will contribute to human and non-human extinction. The play therefore ignites existent enviro-political tensions embedded in contemporary Australian consciousness. Each of the characters delights in their own hubris. Most of them delight in their superiority over non-human nature. Hubris allows them to patronize, neglect, ignore and even abuse that which is other-than-human and humanity's triumph over the natural world.

The play embraces the idea that while climate change and global warming are deeply disturbing, they are phenomena that are absorbed and interred in the usual and everyday—simultaneously unimaginable, almost inconceivable but ultimately inconvenient to the lived experience. For the characters, global warming is an abstract idea, a planetary event that clearly affects the everyday world but ultimately, something that does not *really* threaten their moneyed, triple-glazed, sanitized and filtered stronghold-like environment:

VISI: ... This house is like a fortress.
BASRA: Yeah right. It's just missing the cannons.
VISI: Who's the enemy?
BASRA: Not 'who'. 'What'. Check out the windows. Triple-glazed.
VISI looks at the windows – huge floor-to-ceiling jobs that 'separate' audience and actors.
VISI: Wow. Is this like ... bulletproofing?
BASRA: Weatherproofing. An impenetrable membrane between us and the rest of the planet. We have our own microclimate. You wouldn't know the Anthropocene is taking place outside.
VISI: The what?

BASRA: Climate change. Human generated.
VISI: -
BASRA: You've seen what's happening in Melbourne. I don't need to lecture you about rising sea levels.
VISI: I'm not political.
BASRA: But you're alive. Living in time and place.
A silence. (Carleton, 2016, p. 9)

VISI and BASRA, both environmental activists in their own unique ways, hold firm to cultural narratives that on some level, understand scientific evidence but still find climate change an abstract idea—something to protest against. However, the sceneographic environment in which they are situated, anticipates the dystopian milieu predicting the impending apocalypse. The circumstance contains dark Romantic elements wherein the Gothic sensibility frames VISI and BASRA's human fallibility. Themes of self-destruction, judgement and punishment, guilt and sin are evident.

Viewing the country from an almost dystopian future, the dramatic work attends to the phenomenology of place and experience, to expose imagined consequences of our present disconnectedness to our ecological responsibilities. Suggesting ideological and philosophical tensions associated with rising ecological consciousness and environmental cultural practice, the play pays as much attention to the user (or abuser) of non-human nature as on nature itself. Carleton's deft handling of plot, circumstance and character poses a sophisticated Climate Gothic theatricality. Drawing on his Australian Gothic prowess, his exploration of the environmental crisis is contained in dual perspectives, which align with Dirlick's (2003) theories on how capitalist practices draw renewed vitality from the globalizing consumerist processes on which we are reliant. The peril of the impending crisis, the significant effect it will have on the population and the devastation it will cause to the country is visible in the hubristic characters' adamant denial of accepting any responsibility for the impending catastrophe:

AUGUSTA: I didn't say the climate isn't changing – we all know that – it's irrefutable
But I refuse to be held responsible – either personally or as a species – for the sun's global warming and cooling cycles (p. 30).

The Turquoise Elephant, like *They Saw a Thylacine* and *The Dead Devils of Cockle Creek*, utilizes climate change apathy as responsible for species annihilation. Looking at extinction on the larger global stage, specific circumstances within the play align with discussions on the burden that culture places on our delicate ecologies as humans lose track of the processes of nature and of each other (Plumwood, 2008). As the characters become more and more out of touch with ecological conditions that support or enable life, more and more unrealistic decisions relating to other-than-human species occur. Carleton draws on the horrors of traditional Gothic and depicts OLYMPIA MACQUAIRE as a voracious voyeur using the global extinction crisis as a source of entertainment. A self-proclaimed tourist of warming events and extinction occasions, she values her opportunities of attending these festivities as a privilege only someone of her class and breeding can afford. When she returns from witnessing the final disappearance of the Musk Ox, she wears 'something endangered' and is carrying a plastic bag dripping blood (Carleton, 2016, p. 15). The bag contains the carcass of the Musk Ox, the last of its species, whose death throes made up the spectator sport from which she has just returned. OLYMPIA is unrepentant about her enjoyment of the event, which reinforces her delusions about the purpose of 'the other' and its life value in the natural ecology of species. She is the rapacious predator proudly admitting, ' … she will now only eat endangered species' (p. 21). Preposterously she pontificates that she is saving the species by eating it and advises BASRA to spice it with a few cloves and slow cook it for the ultimate gastronomical satisfaction. The aghast BASRA claims to be living in a chamber of horrors; however, OLYMPIA, oblivious to the error of her ways, advises that meat consumption is now a supply and demand condition whereby the free market prevails, free from environmental fanatics (pp. 17–21). The Climate Gothic reading sits within OLYMPIA's absurd, yet careless disregard of warming repercussions, and in her lack of remorse about the extinction of species.

Carleton's play is erudite in its ability to explore climate change in a tragi-comic theatrical mode. Like Eugene Ionesco's[7] philosophical and moral statement about the spread of Nazism in the mid-twentieth century, wherein the Rhinoceros became symbols of a movement premised by natural selection and survival of the fittest, an *actual* turquoise elephant becomes the symbol against the spread of environmental terror in the city of Sydney:

> AUGUSTA: […] The strangest thing happened on the drive home. I thought I saw something on the streets. An elephant.

BASRA: An elephant?
AUGUSTA: It was blue.
BASRA: What?
AUGUSTA: Or green. The color in between.
BASRA: You saw a turquoise elephant?
AUGUSTA: I KNOW. It's silly, isn't it? Just for a minute. Out of the corner of my eye. Then when I turned to look at it directly ... it vanished.

Beat

BASRA is away with the pixies. AUGUSTA clicks her fingers in front of BASRA's eyes.

Basra, really what's wrong with you today?

BASRA: It's just that - my blog. My blog is called *The Turquoise Elephant*.
AUGUSTA: Is it?
BASRA: Of course it is. I've told you a dozen times.
AUGUSTA: I suppose I must have known that.
BASRA: It's my work.
AUGUSTA: It's a hobby, Basra.
BASRA: It's what I do. I write about climate change. And turquoise elephants. And the Melbourne floods, and you see one. On the streets. (Carleton, 2016, pp. 6–7)

The turquoise elephant also becomes a crucial metaphor for the global sickly green elephant in the room, the weighty beast that symbolises silence, ignorance and the challenges of climate change that are silenced. While symbolizing aggressive or conflicting responses to climate change, it is also the symbol for an environmental activist blog posted by BASRA. As BASRA explains, 'turquoise is the color of the post-Green movement' (p. 11). BASRA believes that the country has breached its environmental tipping point, a concern also offered by a masked CFFEP figure at the opening of the play:

MASKED FIGURE: We are The Cultural Front for Environmental Preservation. We are

> A political art movement who decry politicians and artists who say nothing about the survival of the planet when all around us sea levels are rising, ice is melting, forests are drying up and species are dying on a colossal scale. (Carleton, 2016, p. 1)

Perhaps one of the most disturbing and unique aspects of the work lies in the thematic exploration of the activities of this eco-terrorist arts organization, The Cultural Front for Environmental Preservation (CFFEP). The CFFEP is a radical, violent and destructive organization, committing crude acts on political targets for environmental motives. Used effectively to drive the dramatic plot, the activists involved sit within the tensions between environmental sensibilities and eco-fanaticism as described by Neville and Smythe (2009, p. 49) as the 'seemingly intractable conflict between [human] energy needs and [non-human] environmental concerns'. Carleton utilizes CFFEP as a functional dramatic device to propose how the critical condition of the environment becomes an ecological weapon to facilitate attacks against an unsympathetic public. What establishes Carleton's exploration of the eco-terrorist movement as unique is that the CFFEP activities conducted are framed under the guise of environmental public art, e.g., the Environmental Minister's car, filled with paint, blows up in a spectacularly violent yet colourful event. The outcome of the explosion is a public installation artwork—*Melting Poles*—a Pollock[8] like abstract. The CFFEP's artistic activities as detailed in the opening moments of the play set the tone that underpins the plot and for the theatrical environmental spectacle embedded in the work:

> [*Video of a FIGURE in an ornate decorated mask, voice moderated slightly for the purposes of disguise.*]
>
> MASKED FIGURE: You've heard the overnight news. Sea water has entered Melbourne's sewerage system. The city has awoken to discover its underground rail is flooding. "Oh dear." "We thought this was only supposed to happen in Tuvalu." Meanwhile the World Business Summit convenes in Sydney. Really? Melbourne is quite literally drowning in its own shit, and all Sydney can do is call a Business Summit? Ironically enough, bullshit has been established as the currency of the talks. In response, we, The Cultural Front for Environmental Preservation, have created an ... excremental artwork. We call it: "Cut the Crap – or Have it Back". We hope you enjoy it as much as we enjoyed making it. (Carleton, 2016, p. 1)

We learn that the target of the excremental art activism is the Australian Summit Delegation (led by AUGUSTA) who are covered head to toe in 'shit' because of their inattention and inaction concerning the

Melbourne sewerage crisis and the ensuing panic about fecal-related disease (Carleton, 2016, p. 2). The art works activated within the play embrace a sophisticated contemporary reimagining of Romantic expression concerning art and non-human nature. The excremental installation in many ways pays tribute to the highly stylized impulses of the Romantic tendency to avoid the classical rules of the unattainable. As did the Romantic artists before them, The CFFEP has developed precise techniques of art making. The CFFEP uses their techniques to question climate inaction and they do so with ludicrous, yet effective, results. AUGUSTA who was a recipient of the excremental statement bemoans:

AUGUSTA: People actually applauded. (Carleton, 2016, p. 3).

Excrement as used throughout the play is both a symbol of ineffective climate mitigation and an actual threat to the future of one of Australia's major capital cities. JEFF warns us that we 'can't believe millions won't die [of cholera] just because [it's] a first world city' (p. 30). In the play, the Melbourne population referred to is in chaos, panic ensues as a mass evacuation out of the city is in progress. The circumstance is unsettling as it asks us to consider the likelihood of such an event occurring in one of our major world cities, and what is most disconcerting about *The Turquoise Elephant* is its realistic premise of such a circumstance actually happening. Carleton's proposition is somewhat prophetic as it flirts with a degree of scientific speculation about the extent, speed and impact the continuing rate of rising sea levels of 30 cm or more will have on the Australian coastline by 2050 (Lane, 2020; Maxwell, 2009). The actual risk, though met with some speculation, denial and even irony by the Government and the general public, does suggest a critical threat possibility to the infrastructure of the city of Melbourne. Further, Melbourne authorities are under advice that significant adaptation is required to offset predictions about a catastrophic future flood event. With eight out of every ten Australians living within 50 kms off the Nation's coast, the rising sea level crisis will, as reported in a recent edition of *The Conversation,* require people to adopt new social, settlement and subsistence strategies (Ulm, Williams, Turney, & Lewis, 2018). In the play, Carleton offers a tongue-in-cheek solution by introducing a land development model of New Eden, a new resettlement community. Transitioning to living life in a new climate-affected Australia, New Eden is the entrepreneurial JEFF's vision for a relocation opportunity for selected Australians, a community that is determined by principles of natural selection:

JEFF: The great cull has begun, VISI. We're about to see mass extinction the likes of which you can't imagine BASRA – how many die in Melbourne if the cholera outbreak takes hold. (p. 30)

JEFF in his dubious attempts to get this venture off the ground, build the domed, air and water filtrated colony of New Eden, is trying to con AUGUSTA, OLYMPIA and BASRA out of their fortune. The development and construction of the futuristic environmentally friendly Shangri-La is on land purportedly *donated* by First Nations Peoples, who have, further to their generous donation, agreed to work on Eden's construction free of charge, in exchange for their right to live there:

JEFF: We've made arrangements with the First Nations Peoples in the area. The one's who's land has been donated. We'll provide sanctuary to any of them who donate their labor to aid in the construction.
VISI: You won't pay them?
JEFF: We're providing something more valuable than cash.
VISI: You've taken their land –
JEFF: Been given.
VISI: Taken their land and now you're exploiting their labor?
BASRA: You never said anything about taking Indigenous land.
JEFF: There's no Indigenous or non-Indigenous anymore.
VISI: Yes, there fuckin' is.
JEFF: We're stepping outside of the capitalist paradigm.
OLYMPIA: Well, that's clever isn't it?
AUGUSTA: Genius. (p. 47)

The play incorporates the full gamut of the Climate Gothic grotesque in its tone, which salutes the monstrous behaviours of those who will continue to appropriate the land even while they are indifferent to taking care of it. Carleton's Gothic lens clearly reflects the continuing debates concerning colonization, institutionalized racism, labour inequity and the subsequent impacts on contemporary (post)colonial thought. His salute to the Gothic legacy of colonial processes resonates firmly throughout the plot, e.g.:

AUGUSTA: I have … vague memories of being a pioneer. Arriving on a boat. In another lifetime. With pine tree on the beach. It's all … just beyond the frontier of consciousness. It comes to me in the silent hours. Twilight. And Dawn. (p. 45)

AUGUSTA is referring to the Macquarie's arrival on the shores of Sydney Cove. Her vague memory of arrival in a land believed to be uninhabited and ripe for imperial expansion is one that has informed the construction of the white nation space, which offers the contested themes evident in generating the genre of Australian Gothic drama.

When delivering her Treaty concerning fossil fuel resumption at the WBS, AUGUSTA is the victim of an eco-terror attack, initiated by her domestic helper VISI, a member of CFFEP. VISI's family has entered the country on an environmental resettlement Visa and her experiences as an environmental refugee encourage her in undertaking a life-path as an eco-terrorist and environmental suicide bomber. As AUGUSTA recovers, the visions of the settler past plague her during her convalescence. The post-eco-terror attack surgery has left her as a Frankenstein composite, a monstrous human jigsaw made up of three other people's organs. AUGUSTA's recovery as described instils the Romanticist critique of scientific and technological advancement. While highlighting the value of empiricism, rational thought and the ideal of advancement and progress, AUGUSTA's miraculous survival is morally questionable, attributed to macabre scientific advancements in multiple donor organ replacement:

> BASRA: It's touch and go. They think they might be able to source most of the organs from one person. She's on the top of the list. (p. 38)

Climate Gothic plays such as this one present their own unique challenges when considering the culturally complex content generated from scientific referents. In presenting the implications of scientific advancement, in this case, medical advancement as fiction, Trexlar and Johns-Purta (2011, p. 185) suggest that writers must also contemplate the monstrous 'immensity of scale, both spatially (as a global event) and temporally (as an unprecedented and imminent crisis in human history)'. AUGUSTA miraculously survived the explosion. She is a jigsaw, pieced back together with transplanted organs derived from multi-donors whose memories (it is assumed) filter through her consciousness. The dramatic irony of the situation nominates scientific advancement as a specifically human concern with virtually no progress made in advancing the care for non-human species survival (as evidenced by the Musk Ox incident.) Carleton re-introduces the newly improved AUGUSTA as a particularly grotesque representation of her former absurd, over-stylized self. Her new self enters the triple-glazed compound. Doors are blown open and she hurtles into the room, bandaged from head to toe. A pair of large, dark sunglasses are all that we see of her face. The same absurdly towering bouffant sticks out of the top of the bandages:

> AUGUSTA: It feels like there's another person living inside my body.
> JEFF: Apparently there are three. Donors, that is. Vying for internal supremacy.
> AUGUSTA: I'm a monster.
> JEFF: No ...
> AUGUSTA: [looking out the window] So much humanity. Dazed. Confused.
> JEFF: Heat waves and pollution as far as the eye can see. Not in New Eden though (p. 44).

Climate Gothic themes are viewed as, in and of the cultural processes of scientific progress responsible for human survival in a world of escalating heat—the outside temperature is 48 degrees Celcius—in a world suffocating in its own excretal pollution and ultimately in a world that focusses only on human material processes and survival of the human species.

In drawing our attention to possible future environmental issues, Carleton utilizes his dramaturgical prowess to expose the deficiencies of our current environmental practices. He reveals the laziness of our environmental sensibilities while simultaneously asking his audiences to consider alternative ways of thinking about their relationship with the environment. *The Turquoise Elephant* provokes a range of dramatic responses to nature, climate and habitat. It explores the relationship between climate change and humanity in environmental, social and psychological terms and brings to the forefront the fact that climate change is not only out there on the edge of selfish conscious thought, but is a beast that appears in our immediate inner and outer lives. Carleton ends the play with the destruction of two of Australia's most recognizable and popular icons, the sinking of the Sydney Harbor Bridge and the slow submersion of the Sydney Opera House into Sydney Harbor. Both iconic Australian symbols are a testament to Australian life in the national psyche and as life as we know it submerges into the rising sea, the CFFEP engages in a concentrated attack on power stations, transmitters, banks, big businesses and everything that has corrupted and/or polluted Australia. However, like the disappearance of Sphinx and the final melting of the ice caps in Greenland, the Australian event becomes a spectator sport, and the resolution of the play suggests the characters, though faced with imminent extinction, have learned nothing:

> OLYMPIA: The Opera House is going under.
> Disappearing forever.
> During the Storm.
> Isn't that exciting?
> Will you sit with me and watch it Basra dear?
> [...]
> Oh, silly me. You never go out do you?
> Not for years. Not since you were a little girl.
> You're a watcher. Like me.
> We're watchers, you and I.
> Oh! There! Look! You can see it coming up now!
> The water.
> It's on the concourse.
> Where's the Harbor Bridge gone?
> [...]
> Get a chair and sit next to me.
> This is going to be the best one yet!
> Better than Melbourne.
> Better than Kilimanjaro.
> This will even be better than the Sphinx.
> [*As Olympia peers at the carnage through her lorgnette, the turquoise elephant sheds a tear and its image fills the room…THE END*]. (Carleton, 2016. p. 70)

The Turquoise Elephant successfully engages with contemporary eco-critical discourse and politicized debates surrounding global warming, green business practices and eco-activism. The play is a powerful example of Climate Gothic and a worthy example of how drama is leading the discussion in new contemporary Australian Climate Gothic analysis. It directs its somewhat horrified and disconcerted audience to view unimagined horrors upon a realistically crafted dramatic world. Within the reasonably imagined near future, it places our tensions and our concerns, as well as our apathetic attitude about the speedy escalation of Australian climate events, under a Climate Gothic microscope. The play asks audiences to reflect on, even question their perspectives on what a severe climate-affected future Australia might be like. It challenges us to image the potential horrors that climate change will bring if we don't shift our perception about the other-than-human world. In bringing to the fore questions pertaining to the future ecology of the nation, it integrates eco-critical theory and engages with current theories pertaining to cultural studies of place, history and cultural geography. The Climate Gothic reading is relevant to cultural scrutiny as it dramaturgically defines familiar spatial dimensions in its climate-scape analysis. Having been professionally produced[9] twice and having won the prestigious Griffith Theatre Award, the work has engaged communities in what can be described as a theatrical encounter between culture, people, place and space. The play is a powerful example of how playwriting and theatre production can provoke our ecological sensibilities. In offering a Climate Gothic translation of meaning across perceived space and time—near future, far future and past and present—the play makes a provocative statement about how we—humans—continue to disrupt non-human nature and ignore the climate crisis in this country.

Notes

1. See Hassall (2017) *Contemporary Theatrical Landscapes: The Legacy of Romanticism in two examples of contemporary Australian Gothic drama* which analyzes Romantic legacies in *Dust* (2016) and *Salvation (2013.)* As viewed in these plays, the analysis suggests that notions of *evil* (in the Romantic sense) are in the ecologies of landscape, place and space from which humans are becoming alienated.
2. See a comprehensive discussion on this point in Heddon, D., & Mackey, S. (2012) Environmentalism, performance and applications: Uncertainties and emancipations in *RIDE: The Journal of Applied Theatre and Performance, 17*(2).
3. Spiritual connection to non-human nature in the sense discussed here is associated with contemporary capitalist and materialistic society's alienation from non-human nature.
4. Major-General Lachlan Macquarie took control of the penal colony of Botany Bay in 1809, arriving on the shores with his own regiment of soldiers of the 73rd Battalion. His reforms of the brutal punishment practices, living conditions and well- being of the convicts were formative.
5. A traditional colloquialism used to allude to wool as the source of Australia's national prosperity. A popular expression that acknowledges that in much of Australia's recent history, wool has been the basis of the national economy and the country's major export.
6. Christopher Charles Skase (18 September 1948 – 5 August 2001) was an Australian businessman who later became one of his country's most wanted fugitives, after his business empire crashed spectacularly and he fled

to Majorca, Spain. Greg Ansley in an article in NZ Herald in 2005 stated that Skase was reviled not because of the hundreds of millions of dollars he cost his shareholders—Australia was littered with such failures—but because he refused to take his punishment as others had, because he never once showed remorse, and, worst, because he laughed in Australia's face.

7. *Rhinoceros* (1959) informs part of Martin Esslin's study of post-war avant-garde drama *The Theatre of the Absurd*.
8. Jackson *Pollock* was an influential American painter, and the leading force behind the abstract expressionist movement in the *art* world.
9. *The Turquoise Elephant* has had productions at Griffin Theatre Company, Sydney, 14 October – 26 November 2016 and Browns Mart Theatre in Darwin 8–19 May 2018.

References

Carleton, S. (2016). *The turquoise elephant*. Playlab Press.

Dirlick, A. (2003). Global modernity: Modernity in an age of global capitalism. *European Journal of Social Theory, 6*(3), 275–292.

Glotfelty, C. (2006). Introduction: Literary studies in an age of environmental crisis. In C. Glotfelty & H. Fromm (Eds.), *The ecocriticism reader: Landmarks in literary ecology* (pp. 161–175). University of Georgia Press.

Hassall, L. (2017). Contemporary theatrical landscapes: The legacy of Romanticism in two examples of contemporary Australian gothic drama. In S. Green & P. Hetherington (Eds.), *TEXT Special Issue Website Series: Romanticism and contemporary writing – legacies and resistances*, *41*, pp. 1–14, October. textjournal.com.au/speciss/issue41/Hassall.pdf

Hassall, L. (2020). Contemporary performance and climate change: Redefining the Australian landscape narrative. *Australasian Drama Studies (ADS), 76*, 270–295.

Heise, U. K. (2008). *Sense of place and sense of planet: The environmental imagination of the global*. Oxford University Press.

Lane, I. (2020). Climate change cities index: Melbourne predicted to be among worst affected by 2050. *The New Daily On-line*. https://thenewdaily.com.au/life/science/environment/2020/01/29/climate-change-cities-index-2050/

Jordan, D. (2014). *Climate change narratives in Australian fiction*. Lambert Academic Publishing. https://doi.org/10.1002/wcc.105

Maxwell, A. (2009). Postcolonial criticism and ecocriticism and climate change: A tale of Melbourne underwater in 2035. *Journal of Postcolonial Writing, 45*, 15–26.

Neville, K., & Smythe, L. (2009). *Environmental activism or national security threat? Policy options for addressing radical environmental targeting*. (Centre

for International Relations Working Paper No. 49, pp. 1–21). http://citeseerx.ist.psu.edu/viewdoc/summary?doi=10.1.1.532.3204
Picart, J. C., & Greek, C. (2010). *Monsters among us: Toward a gothic criminology*. Teaneck Fairleigh Dickinson University Press.
Plumwood, V. (2008). Shadow places and the politics of dwelling. *Australian Humanities Review, 44,* 139–150. http://australianhumanitiesreview.org/2008/03/01/shadow-places-and-the-politics-of-dwelling/
Trexlar, A., & Johns-Purta, A. (2011). Climate change in literature and literary criticism. *WIREs Climate Change, 2*(2), 185–200. https://doi.org/101002/wcc.105.
Ulm, S., Williams, S., Turney, C., & Lewis, S. (2018). How sea levels could shrink Australia and spark a coastal exodus. *The Conversation: ABC News on-line.* https://www.abc.net.au/news/2018-01-16/rising-sea-levels-could-shrink-australia-coastal-exodus/9333400. Accessed on 28 July 2020.

CHAPTER 7

Greening Theatre Landscapes—Sustainable Australian Theatre Production

Abstract As sections of the population become increasingly aware of the threat of environmental degradation, more theatre professionals and organizations are becoming concerned about the particular negative impacts (i.e., the carbon footprint) theatre productions leave. In recognizing, the theatre industry's response to climate change and its potential to encourage change in climate behaviours, it would be hypocritical perhaps to disregard how applying 'green' practices in producing theatre is way of promoting climate action. This chapter therefore offers a discussion on 'green' production practice and draws on 'green' initiatives from an Australian theatre industry perspective to support the discussion. Many Australian theatre artists are implementing 'green' processes and encouraging sustainable practice futures in the performing arts. This chapter discusses how theatre can promote sustainable production practices. The chapter refers to organizational sustainable production processes and while furthering discussions on the Climate Gothic from an artistic perspective, it examines the sustainable production processes surrounding a unique Climate Gothic play, *Water Wars* (2011) by Elaine Acworth.

Keywords Green Theatre · Sustainable Production · Eco-Theatre

L. Hassall, *Theatres of Dust*,
https://doi.org/10.1007/978-981-16-6159-4_7

This is an important addition to the discussion on Climate Gothic as explored in other chapters. The discussion acknowledges that theatre production traditionally, and in many cases continues to be, resource voracious, especially in relation to wastage—both energy and resource wastage in design and administrative areas. While the plays discussed in this book provide an important discussion on the development of Climate Gothic as a burgeoning genre responding to the global crisis, the very real, emotional, ethical and reasonable challenges theatre makers face require further consideration. It has been a generational challenge with sustainable production applications in Australian theatre, stemming from what Beer and Hes define (2017) as an industry that has traditionally been dominated by suppositions and speculation that 'green' means limiting aesthetic design frameworks.

Theatre productions' unsustainable practices are problematic and have been for many years. The amount of waste produced in scenic construction and marketing and administrative processes is historically excessive. Energy wastage through inefficient lighting design is exorbitant. A consistent problem with theatre resource waste is associated with traditional popular twentieth-century big-budget aesthetics concerned with theatrical spectacle, predominantly promoted by Entertainment Centers and State Theatre, Opera and Ballet organizations. Large elaborate sets, extravagant lighting designs and even excessive marketing wastage in terms of posters, flyers and programmes have contributed to theatre wastage from energy and resource perspectives. Fried and May noted theatre's unsustainable practices as far back as 1994 and claimed, 'The amount of waste produced in scenic construction is prodigious' (p. 28). Much later Morris (2007) echoed this criticism describing the theatre industry as '… a business of waste, [that] reflects a consumptive society' (p. 18). This criticism has considered how it is easier to continue traditional, though unsustainable, processes as the challenges posed by researching, sourcing, and then implementing green technologies or developments can interrupt the continuation of a status quo, e.g., big budget productions are acknowledged as 'better' or 'more professional'. Alternatively, smaller independent productions may prove inadvertent frontrunners in sustainable practices due to necessity, e.g., they may be greener in overall outcomes due to limited budgets and the programming of productions in smaller venues. Lack of attention to sustainability considerations might also have to do with the fact that greener resources suitable for theatre production needs are often more expensive as they are new and/or developing technologies. In 2008, Robert Butler from The Centre for Sustainable Practice in the Arts alerted us to the fact that unsustainable theatre production wastage is an endemic theatrical hole in which most theatre artists and organizations fall into and offer numerous reasons why this is so:

- Climate issues in the theatre do not easily fit theatre's cause and effect dramaturgical model.
- Science suggests technical/factual concerns that may prove alienating to the imaginative freedom of theatre's artistic philosophies.
- Difficulties and costs associated with changing organizational excessive energy practices and processes that may affect ways in which practitioners are used to operating within the organization.
- Theatre-makers are unsure of their own stance or response to global warming and climate issues.
- Companies think going green will affect programming (and as a result audience attendance) by assuming this may include apocalyptic or dystopian stories.
- Provocative or agitational ecological or environmental issues may be at odds with sponsors e.g. fossil fuel companies (Butler, 2008).

As these issues continue to challenge us, there is no argument that theatre production has been slow on the green uptake. It was evident from as early as the 1990s that global warming was a significant challenge requiring a comprehensive human paradigm shift to address, with that year's Greenpeace report warning:

> The uniquely frustrating thing about global warming – to those many people who now see the dangers – is that the solutions are obvious. However, there is no denying that enacting them will require paradigm-shifts in human behavior … There is no single issue in contemporary human affairs that is of greater importance. (Leggett, 1990, p. 457)

The theatre industry and its practices are not exempt from sustainability rationalization in such a paradigm shift. Applying green production methods in the theatre industry is one way theatre makers are addressing this shift, with many theatre and performing artists and organizations declaring their commitment to reducing theatre's carbon footprint in this country. As such, the Australian theatre industry is slowly developing green(er) production philosophies and recognizing the need to promote sustainable processes and practices. In 2008, Queensland Theatre Company[1] joined forces with URL (an environmental global engineering firm) to establish the Green initiative to improve sustainable practices across both organizations. In 2016, Queensland Theatre Company further committed to improving its sustainability profile and embarked on engaging with the Department of Public Work to ensure the achieving of best possible greening outcomes. Many high-profile professionals such as Michael Kantor, former Artistic Director of Melbourne's Malthouse Theatre,[2] reviewed the 'business as usual' model of theatre production within Australia:

> We know the planet is facing an environmental catastrophe of a scale and impact unlike anything it has faced in the period of human civilization. Yet we continue to develop and produce our theatre, in our energy intensive theatres, in our energy intensive ways, to our fickle and dwindling audiences, as if we didn't know of this impending catastrophe … business as usual. If there is one thing that is clear … it is that [going green] requires changes to every part of our lives, as individuals responsible for our own carbon footprint, and as participants in a 'culture industry'. (Kantor, 2010, p. 10)

Going green in theatre practice suggests a transformational change that requires a shift in thinking about how theatre as a business behaves sustainably and consequently how others perceive the behaviour of that business. Many organizations and individuals took up the challenge. Sydney Theatre Company won the Green Globe Award in 2010 for Sustainability Excellence with its 'Greening the Wharf' project, which transformed the beautiful heritage-listed Sydney Harbor wharf, and buildings into an inspirational example of environmental theatre company leadership. In 2016, Opera Queensland toured a green(er) production of the *Mikado*[3] but also acknowledged that there is a significant need for Australia to collect and make available suitable benchmarking statistics for industry use about greening touring productions. In 2016, a report in Greener Live Performances, an industry watchdog, identified that in replacing old technologies with energy-efficient LED (Light Emitting Diode) lighting, the Perth Concert Hall and the Melbourne Recital Hall attained sustainability targets in their productions. The Malthouse Theatre has gone completely rechargeable in battery usage for sound equipment. Regional venues such as the Mackay Entertainment and Convention Centre (Queensland) are more energy efficient through the implantation of LED lighting technologies in production and in public areas such as the foyer. In 2013, The Australian Major Performing Arts Group (AMPAG) reported that The Australian Ballet estimates that the introduction of green policies and subsequent practices is saving the organization approximately $10,000 per year. The sustainable and green production practices implemented across these major performing arts organizations required the attention, involvement and collaboration of artistic personal, design and technical production artists and venue administration and management. Through team and managerial collaboration, the organizations were able to maximize improvements in theatre's carbon footprint through their day-to-day management of production processes, practices and behaviours. Apart from major organizations such as those listed above, many independent artists and scholars in higher education institutions are also leading by example[4] and becoming conscious of the ecological stewardship associated with producing sustainable theatre. Australian designers are at the forefront of taking on the responsibility to be industry leaders and educators in sustainable practices, and their knowledge of environmental and ecological practices is evidenced in the work of scenographers such as Penny Challen and Tanja Beer. In promoting sustainable practices and processes, the Australian industry is showcasing that theatre artists are (a) aware of the global environmental problems associated with global warming and climate change and (b) are transferring this knowledge to production processes that have inherently been wasteful. As such, it is encouraging to note that while there is a long way to go, Australian theatre makers are rethinking resource usage and the way we make theatre in the Anthropocene.

Why is sustainability so important in the theatre industry? Is there significant aesthetic impact on the development and production of a theatre production when adopting a sustainability ethos throughout a project? Beer and Hes (2017) suggest it is a fallacy to claim that green practices limit aesthetic outcomes, and discuss how 'contemporary ecological concerns bring with them opportunities' for cognitive, practice and creative innovation, and provide theatre artists with a way to 'push intellectual and creative boundaries' (p. 32). The play discussed in this chapter is an example of how artists are, as described so eloquently by Alison Tickle (2016), putting 'sustainability and its principles of resilience and strength, creative regeneration and respect for the earth at the center of [their] artistic vision'. *Water Wars* (2011)[5] by Queensland playwright Elaine Acworth, produced by Umber Productions at Oakey Entertainment Centre and La Boite Roundhouse Theatre, Brisbane, is a play about sustainability. Produced sustainably, the play responded to the long-term drought that affected Australian lifestyles between 2003 and 2007. During this time, some regions in Queensland were subject to the toughest water restrictions in place in Australia of all time. The highest level of water restrictions reached during the Queensland drought was Stage 7 for Brisbane, and South East Queensland reached Level 6 restrictions. Overpopulation, the drought, coupled with corresponding reductions in the supply of drinking water, led Australian State governments to consider alternative water sources such as recycling wastewater to supplement existing sources, which divided an already stressed and water-starved population. At its heart, the play explores how communities divide when facing water shortages and how friends and neighbours turn on each other, quick to report over-usage to the 'water inspectors' deployed by the State governments, who could issue penalties to those seen to be wasting water[6]:

> *Gally stands at the tap – the flooded ground around her. Pat is also staring at the ground. Bernice watches everyone.*
>
> GALLY: It was an anonymous tip on the tip line
> ...
> Someone called and left a tip on the tip-line about me.
> PAT: I said, that's very wet.
> BEREN: Bound to notice.
> GALLY: One day we've been here.
> PAT: Well, some of us're not so neighborly.
> GALLY: They can remotely access meters nowadays; they can tell you to the minute when water's drawn from the mains and when it stops.
> PAT: Can they really?
> GALLY: So when this complaint came in, they read my meter and what did they find? It drew water at midnight AND for 20 minutes from 4am this morning when none of us were up.
> *Silence* (Acworth, 2011, p. 29, La Boite Draft)

Water Wars is a play about how we live with, and survive each other, on an increasingly fragile planet. Written at a time when enforcing excessive water restrictions was the norm, restrictions were placed on showering, watering lawns, using sprinkler systems, washing vehicles, and refilling swimming pools. The play is a testament to the impacts of climate change on everyday behaviours, and relocates the climate-beast into our own suburban backyards. The play can be read and viewed as an environmental fable that responds to these issues, but is set in 2025—the final years of the drought—in what by then has become a police state,[7] where violating the Water Crimes Act incurs harsh penalties.

> GALLY: And there's notice to show cause within 7 days at the Water Courts as to why we should not have out town water turned off and be given only an allocation of 30 litres per person, per day delivered in a tank – there's that Pat! ... 30 litres ... I've got to keep the trees alive for Cal; I can't possibly do it, I can't possibly keep us and the trees alive on 30 litres a day ... (Acworth, 2011, p. 49, La Boite Draft)

As the title suggests, central to the *Water Wars* story is the theme of water, its usage, conservation, control and wastage. Through the conflict between three sets of neighbours, the play explores themes of what we, as a society, may face in the future about water consumption, regulations and restrictions. The premise of the play responds to the actual restrictions as described above that Australians were required to comply with during the lengthy drought. At the outset of the play, we are privy to the fact that water is scarce, so scarce that the government has imposed strict water restrictions with exorbitant fines for water wastage. The plot follows the neighborhood community in the final drought year, at odds with each other over Level 5 water restrictions. The characters are all flawed, each, and every one of them, greedy for access to the water for purposes other than survival, but which they desperately believe they need:

> ROBERT: It's gone! The water is gone.
> BEREN: What do you mean, gone? You mean – not there?
> ROBERT: Gone, Doll. The tanks are empty.
> BEREN: Well, get more; fill up again!
> ROBERT: Listen. We bought two fucking big tanks worth 18months ago when we had some money: 250 000 Litres that cost an arm and a leg. Now, I've sold the lot of it on the black market for less than I paid and I've spent the lot on bridging loans – we're in deep shit baby –
> *Silence as she considers this.*
> BEREN: But how will I get my pool if there's no water?
> ROBERT: I don't give a flying fuck about your pool!
> BEREN: But I need my pool ...

ROBERT: Alright, alright, alright ...
BEREN: You think you are something then it all gets taken off you – and they don't even mention your name ...
ROBERT: Christ's sake: what do you want, Doll?
BEREN: More.
ROBERT: More?
BEREN: More! My dues! I want a pool across two levels that plays as a waterfall down a shining rock face and seems to end in the distance in a little strip of golden sand; with tall reed beds, which run gently down the hill beside it, right there. That's me; that's who I am; you have to give it to me, save me, you do that, you give me, you make me happy.

She has collapsed on the ground.

ROBERT: Doll ...
BEREN: That's what you DO, Bobby, you give me –
ROBERT: All right, wait a minute; well ... maybe... (Acworth, 2011, p. 45, La Boite Draft)

The characters slowly descend into a form of suburban madness as their greed for water becomes all-consuming as the restrictions escalate. A suburban water war ensues as the characters attempt to survive both the restrictions and each other from their own egocentric perspectives. This play itself is a significant testament to the Climate Gothic as described in this book. The climate-scape is inner-city Brisbane and offers an urban Gothic futuristic examination on what has become a Queensland (and Australian) backyard Grimm-like fairytale. The Australian dream—tending green sprinklered lawns, nurturing lushly flowered gardens and raising well-watered vegetables and after the gardening is done, sprawling lazily under huge jacaranda or mango trees to then frolic at will in cool inviting swimming pools—is fractured, ruptured through the reality of double perch[8] parched yards, empty pools and dying trees:

Scene1: Day 1, late afternoon.

Gally's backyard. Cal scrambles back up into the treehouse, Gally buckets to the yard from the washing machine. The sound of the machine emptying and filling another bucket.

GALLY: There'll be a forest here, one day, Cal – your forest, your trees. A patch of backyard green – one among many, maybe, hey?

> Imagine a whole lot of yards, all joined up by trees. All the eye sees is green.
>
> *Gally carefully waters one of the half dozen saplings in the yard* (Acworth, 2011, p. 7, La Boite Draft).

The hope for a forest of trees in CAL's own back yard, a dry and drought affected urban climate-scape, draws on nostalgic memory in tone. The lushly treed landscapes that once defined the Australian backyard is a wish. In its incorporation of themes of loss, nostalgia and memory for disappearing non-human nature, the play situates itself as a significant example of the Climate Gothic. *Water Wars* significantly contributes to Australian climate change discourse in the cultural realm. However, what is unique about this specific work is that its production process was almost entirely sustainable. *Water Wars*, at the time of production, boasted a distinctive process that agitated for sustainability both creatively and practically, testifying to its contribution as a strong example of the way artists are applying green theatre philosophies and technologies in contemporary theatre making processes.

It is encouraging to note that Australian theatre industry makers are no longer hiding behind thinly veiled excuses that the small amount of damage done by one show or one company in one small country is insignificant when compared with the magnitude of the problem globally. Clearly, Umber Productions worked on the principle that the development and production of *Water Wars* actually did count in the larger scheme of things. By questioning its organizational and production principles from a green philosophical perspective, the company and the artists involved in the production made a conscious shift to sustainability choices. While traditionally the consensus suggested that thinking and acting ecologically is risky due to perceived aesthetic limitations (Beer & Hes, 2017; Fried & May, 1994; Rowan, 2015; Rowan & Hassall, 2019), from its inception, ecological issues pertaining to climate change, drought and water conservation were at the foreground of the aesthetic imaginings of *Water Wars*. The adopting of a sustainability ethos to produce the play framed the early rehearsal and production process. In *Water Wars*, themes of water conservation and sustainability led to a theatrical process that incorporated sustainability production points of view from creative development to production. In discussing and debating sustainable practice and process ideas for the production, it was evident that the themes drawn from the ecological, philosophical underpinning of the text were of the utmost importance to the decision-making process. Umber Productions focused on developing and implementing a sustainable philosophy that was particularly effective in areas of set design and construction and lighting design and application. Lighting designer David Walters (2010), in an on-line discussion of the project about the green lighting design, stated:

> ... the premise of play is that it is [an] 'environmentally friendly touring package ... we [will] use power very carefully in the production and try and produce a piece that is...ecologically friendly.

The project's original designer Greg Clarke (2010) stated in the same presentation:

> We've definitely got an over-arching [design] aim ... we're ... approaching a production (design) that philosophically and practically lives up to the themes of the play, so it is something that does use low energy, does conserve energy ... which is a very exciting, very new idea for me.

From the *Water Wars* perspective, the ecological thinking and processes framing the production generated new ways of thinking about possibilities for theatre production from the company's perspective. When considering the lighting design aspect of *Water Wars* through the lens of ecology, the notion of electrical power consumption came into focus as one of the strongest recurring themes to promoting production sustainability. In any production, energy consumption (mostly old technology that is slowly being superseded with LED alternatives) is almost exclusively related to the use of theatrical lanterns—the lights used in a production to light the stage, light the actors and assist in creating mood and atmosphere to support the director's artistic vision. The *Water Wars* design process was unique at the time of its production season (and in many ways still is), as the production team practiced what the play preached in terms of energy consumption. As a result, the main focus of the production processes was that of electrical energy consumption—particularly in the area of stage lighting and the choices of materials used and processes of set construction, as well as an appendage supporting the production—the adoption of e-marketing.[9] What the design team achieved was the ability to create a design that was consistent with ecological ideals embedded thematically in the text (Rowan, 2015).

The lighting design was a significant inroad into the development of green and sustainable processes in Australian theatre and is a tribute to how a theatre production does not need to generate a huge carbon footprint to be successful. Theatrical lighting technology is changing at a very fast rate and newer, more efficient technologies are increasingly gaining popularity, as intimated earlier. At the time of *Water Wars* in 2011, the greatest challenge in implementing energy efficiency measures in the lighting design related to achieving a balance between visual artistic outcomes and energy consumption. To achieve energy objectives that included a low impact carbon footprint, the lighting designer, David Walters, in association with Umber producers imposed an extraordinary limitation to the design framework. No lighting state within the show would draw more than 10 amps (2400 watts) of electricity—the amount of electricity that is available from a general power outlet (GPO), or domestic power point (the ones used in homes, offices, etc.). To contextualize what this limitation meant, the theatre industry is notorious for energy consumption with typical lighting design processes having scant regard for power consumption with design frameworks more concerned with producing a strong visual atmospheric aesthetic, often framed by hundreds of lighting cues operated by three-phase power. Given that the creative integrity of the *Water Wars* lighting design was essential to the aesthetic, the energy efficiency actions of including an overall power limit on the design was the biggest challenge in the design process for the lighting design team. In terms of energy efficiency, the production did not use any three-phase power typical in lighting design implementation and operation. In its bold decision to draw no more than 10 amps of power at any one time, the entire lighting design ran off one domestic light socket, showcasing how lighting design technologies can be both sustainable and aesthetically pleasing. Further *Water Wars* had over 400 technical cues, yet the maximum power drawn at any one time was 7 amps. The challenges associated with this 10-amp (2400 watts) limit led Walters to conceive unique ways of thinking about lighting. He incorporated a range of low wattage lighting fixtures using LED lanterns including: 60-W tri-colour LED Par 64s, LED Strips, Par 36 pin-spots and 30-W LED profiles. The design also incorporated the use of 12-V incandescent lanterns and battery powered torchlights to achieve the lighting requirements of *Water Wars* (Rowan, 2015, p. 40). This was a significant achievement and further, a significant outcome in green production applications in lighting design in this country.

The sustainability framework for the artistic vision and the production of *Water Wars* included eco-creative design elements that are achievable in professional, independent and community contexts, even at grassroots administrative levels. Change can be applied at various levels and across various areas, and as playwright Elaine Acworth describes,[10] in areas of:

> **Set Portability:** The set of *Water Wars* was designed for easy transportation with touring considerations in mind.[11] It was made from tough recycled card product, was lightweight and able to be constructed easily in pieces. The pieces were digitally printed upon and were used to represent both sceneographic place and provide background for AV projection when place or circumstance shifted.
> **Recycle, Re-purpose, Reuse:** Design items for *Water Wars* such as costumes, water buckets and hoses were borrowed (not bought), used and then returned to owners at the end of the production. If any design item was purpose built or purchased, it was used, then returned to Umber stock items at the end of the Season.
> **Ecologically Sound Suppliers:** Set Design suppliers were chosen who offered energy saving solutions and/or sound environmental credentials. All set and staging were constructed using recycled timbers with fabric items recycled and repurposed.
> **Paperless Marketing:** No printed paper materials were used. Digitized Programs were available for downloading on smartphones or mobile devices.

The producers of Umber Productions collaborated with the designers of *Water Wars* in a proactive, collaborative approach to design aesthetics. Stephen Rowan, the production manager of *Water Wars* claims that the biggest sustainability feature pertaining to the materials that were used in the set construction came from the use of a recycled non-toxic cardboard product and the 'sustainability' aspect of the design revolved around this recycled product traditionally used in the commercial area of trade shows (2015, p. 46). Traditionally, and especially in professional productions, set designers are given carte blanche over choice of materials and processes they choose to incorporate into sceneographic elements (Hassall & Rowan, 2019). The product was a fairly strong and rigid recycled cardboard that could be used to skin flats (covering a timber frame to create a flat but solid surface). The product was effective from sustainable and aesthetic perspectives in that it was 'green' but also effective for projecting scenic environments and locations onto. Rowan (2015) notes that the sustainability ethos framing the *Water Wars* production both informed the construction processes of the design and, importantly, sustainability was at the forefront of considerations for the sustainable recycling or disposal of set pieces. As noted above, the set design elements for *Water Wars* developed through a combination of materials that were either already recycled and/or could be recycled at the end of the season, with these decisions aimed at overcoming waste management issues that are traditionally excessive. The cardboard product sourced for use in the flats of the production was completely recyclable, and at the end of the season were disposed of in the cardboard recycling bin. Essentially there wasn't a single set element that couldn't be recycled, reused or broken down and put through some kind of recycling process at the end of the La Boite Roundhouse Theatre Season. As a result, the production team worked eco-creatively in using materials that served the story best. The set design processes effectively incorporated sustainability considerations, including 'green' applications within the artistic vision, choice and use of materials, construction and disposing of any wastage.

Reduction of waste factored into other areas of the production process, specifically marketing and administration. In order to align with a sustainable ethos, the production team identified the need to limit paper consumption. This was undoubtedly another significant challenge for the team as the myriad administrative tasks required within a theatre production process (including those associated with stage and production management and rehearsal processes) generally require the consumption of a great deal of paper. Marketing processes, including ticketing, posters, flyers and programmes occurred through on-line platforms with programmes being downloadable from patrons' smartphones. Paper use processes were therefore limited, and printing processes were bypassed. Limitations placed on the printing of various drafts of scripts during the creative development and rehearsal process, scheduling, production minutes, action plans, script notations and edits, prompt copy preparation and maintenance lists also occurred. In these areas, the *Water Wars* production team 'greened' the framework around marketing and administration processes. The stage management team attempted to respond to the sustainability ethos of the production in its entirety by utilizing on-line updating of paperwork prior to printing. The production team noted that one of the main challenges in this area was the fact that the script was still being developed and edited during the rehearsal process (Rowan, 2015, p. 48), necessitating the constant updating of developing drafts. However, to avoid unnecessary paper consumption, only newly edited pages were printed. Some of the cast chose to adapt to greener adjustments in the use of paper. Traditionally an actor has a paper copy of the script during rehearsals, using this as a prompt during the rehearsal process, to mark in their blocking (where and when they move upon the acting area), and to mark in directorial notes. With green considerations in mind, however, some actors chose instead to make use of an iPad rather than a paper-based script. While there is now more access to software for more eco-friendly Stage Management processes, at the time, this software was expensive and hard to access, so the challenges addressed and the processes implemented by the Umber team were commendable.

Claudia Orenstein in 1998 argued that no theatre production 'could possibility take responsibility for inspiring apolitical audiences in a complacent era to concrete political action' and that the context of the performance 'determines potency' (p. 6) of the message delivered. It is encouraging to note that now, over two decades later; sustainable production processes are now on the agenda of most theatre organizations and production activity. Climate action is at the forefront of Australian political debate and public response,[12] especially that of the younger generations from which our emerging theatre makers of the future will arise. *Water Wars* is an example of how we can prepare future generations of theatre makers for the concerns of social responsibility, prescribing that attention to global climate change challenges can appear in theatrical methods of delivery. The play acknowledges that if theatre is to function as a social weapon in the education or agitation about climate change with and for general public audiences, it is reasonable to assume that it does need to contain some form of green activism that argues against traditional unsustainable processes. In joining the paradigm shift described in the introduction to this chapter, *Water Wars* is an example of how sustainable theatre production is accomplished and can resonate through theatre communities, activating ecological stewardship in emerging theatre makers. Ecological or sustainable management in theatre production can be a slow process of self-assessment. Assessing big business as usual models of production requires a shift in personal values. Change occurs through considering and developing alternative possibilities, e.g., power, materials and reorganization, and in rethinking aesthetic possibilities. *Water Wars* exemplifies how achieving sustainable Australian theatre includes organizations, artists, people and audiences in a shared dream of an ecological sustainable future for the future benefit of continuing the climate change conversation through the cultural platform of an important industry.

Notes

1. Queensland Theatre Company (QTC) has been recently renamed Queensland Theatre (QT).
2. For over 30 years, Melbourne's Playbox theatre fanned the flames of Melbourne's radical culture, exerting significant influence on the city's artistic community. In 2004, Michael Kantor reimagined Playbox as Malthouse Theatre. Malthouse Theatre collaborates with local and international artists and aims to create inventive performances that cut to the core of the human experience.
3. See http://greener.liveperformance.com.au/case-studies for case study.
4. See Rowan and Hassall *Greening Theatre Landscapes* (2019), which explores a production process with 3rd year contemporary performance students that was implemented through a sustainable production process. The discussion provides insight into how sustainability literacies can be embedded in theatre performance and production practices in a university context and consequently offers a model of sustainable theatre education that may adopted by others in the university and tertiary sector. Further it discusses how ecological stewardship is an important graduate attribute for emerging theatre artists.
5. *Directed by Shaun Charles, (2011, August 3–20 LaBoite Theatre Production Season).*
6. A popular anecdote in Brisbane theatre circles associated with the play is that playwright Elaine Acworth's aging mother was fined for watering her garden and this is what initially sparked the idea for the play.
7. The police state alludes to a government that exercises power through the power of the police force. In *Water Wars*, the term has taken on a derogatory meaning by describing the undesirable state of living with water restrictions and is characterized by the overbearing presence of civil water authorities. The inhabitants' use of water is monitored as severe and punishing restrictions are enforced.
8. Perch refers to a term for an Australian house block of land, usually in older, more affluent suburbs referring to 800 m square.
9. See Stephen Rowan's (2015) research on sustainable theatre production for full discussion on the *Water Wars* processes. Rowan worked as the Production manager on the Touring Pilot Season and on the La Boite

Season. The production inspired him to undertake research into Australian theatre's green practices: *Water Wars:* Exploring the tensions, challenges and possibilities of applying a sustainability framework to the development and production of a new independent theatre performance, Griffith University.

10. See Acworth's on-line notes about Water Wars (https://greener.liveperformance.com.au/uploads/files/M3_GLP_WaterWars_CaseStudy_Final-1413860197.pdf).
11. *Water Wars* had a pilot production at the Oakey Cultural Centre on the Darling Downs in Queensland prior to the La Boite Season. It was supported by The Empire Theatre, Toowoomba. 20 July 2020.
12. In 2019 tens of thousands of young Australians walked out of classrooms to stage protests in all Australian capital cities and in many regional centres across the Nation. They protested against what student leaders described as the destruction of their future. These protests were inspired by the action of Swedish student Greta Thunberg.

References

Australian Major Performing Arts Group. (2013). Greening the performing arts. https://www.ampag.com.au/article/greening-the-performing-arts

Beer, T., & Hes, D. (2017). Sustainability in production: Exploring eco-creativity within the parameters of conventional theatre. *Behind the Scenes: Journal of Theatre Production, 1*(1), 32–54.

Butler, R. (2008). Why theatres don't touch climate change. https://sustainablepractice.org/2008/11/12/why-theatres-dont-touch-climate-change/

Clarke, G. (2010). Water wars creative development ii, Umber Productions Archive Video. https://vimeo.com/14918516

Fried, L., & May, T. M. (1994). *Greening up our houses: A guide to more ecologically sound theatre*. Drama Book Publishers.

Greener Live Performance (2016). Case studies: Energy efficiency measures in live performance venues across Australia. http://greener.liveperformance.com.au/uploads/files/case_study_best_practice_venues-1412900492.pdf

Hassall L., & Rowan S. (2019) Greening theatre landscapes: Developing sustainable practice futures in theatre graduates. In W. Leal Filho & R. Leal-Arcas (Eds.), *University initiatives in climate change mitigation and adaptation* (pp. 143–158). Springer. https://doi.org/10.1007/978-3-319-89590-1_9

Kantor, M. (2010). *Towards a new environmental theatre: An aspirational manifesto for the theatre community*. Tipping Point Australia website: https://static1.squarespace.com/static/583e4a63ff7c50809923ce78/t/5980495f579fb31d6e7151c5/1501579755597/Greening+the+Arts.pdf

Leggett, J. (1990). *Global warming: The Greenpeace report*. Oxford University Press.

Morris, D. G. (2007). *Towards a recycled theatre: Industrial ecology theatrical applications for the next industrial revolution* (Doctoral dissertation, Western Washington University).

Rowan, S. (2015). *Water wars: Exploring the tensions, challenges and possibilities of applying a sustainability framework to the development and production of a new independent theatre performance* (Honors thesis, Griffith University).

Tickle, A. (2016). Sustainability should be at the heart of our collective artistic vision. https://www.theguardian.com/culture-professionals-network/culture-professionals-blog/2012/oct/25/sustainability-arts-council-julies-bicycle

Walters, D. (2010). Water wars creative development ii, Umber Productions Archive Video. https://vimeo.com/14918516

Water Wars by Acworth, E. (Playwright). (2011) [Theatre Production] La Boite Independent Season, in association with Umber Productions, Kelvin Grove, Brisbane City, 3–20 August, 2011.

CHAPTER 8

Conclusion Theatres of Hope

Abstract This chapter provides a final overview of the importance of situating Climate Gothic as a field of drama analysis in contemporary Australian playwriting and performance culture. The concluding discussion suggests that the plays and performances provide audiences of theatre production as well as readers and scholars of dramatic fiction with the opportunity of witnessing, and responding to, cultural change. In doing so, the chapter concludes that the cultural myths concerning the country, which have traditionally informed our understanding of who we are as a people and a Nation, are now in question. Further, the conclusion posits that while the traits promoting our inherent Australian-ness to the rest of the world are changing in alignment with the warming planet, the Climate Gothic plays discussed offer us hope for a better future. The chapter draws on excerpts from *Dust* to conclude the discussion.

Keywords Climate Gothic · Australian Playwrights · Contemporary Culture

The warming of the planet is likely to remain both an urgent crisis demanding attention and mitigation, and a subject of contention. However, it is more than apparent that we can no longer ignore climate

L. Hassall, *Theatres of Dust*,
https://doi.org/10.1007/978-981-16-6159-4_8

change as something that might happen, as something out there, somewhere on the fringes of our consciousness. It is happening now, and we are living it. Climate Gothic plays and the identification of the climate-scapes framing them challenge traditional Australian dramatic functions concerned with establishing our nation and our identity and extend Chaudhuri and Fuchs' (2002) landscape discussions about the role current environmental and spatial experiences have in informing cultural meaning. Engaging in this dialogue is especially relevant in the contemporary Australian context, as an understanding of where we are now in the warming world assists us in engaging with politicized climate-driven debates about our environment and disappearing ecologies.

The escalation of global warming on the landscape underpins new dramatic and theatrical imaginings that aim at formulating some critical perspective on the materialistic cultural mechanisms responsible for affecting the environment. The plays discussed in this book in no way measure the breadth of Climate Gothic work emerging by Australian performance makers and playwrights. The playwrights discussed here are few of the many theatre artists responding to global warming, who are pivotal in paving new climate-scapes in Australian drama through their reimaging of the quintessential bush, desert, beach and outback landscapes. In drawing on the Gothic insights that prescribed the popular Australian drama of the 1990s and early 2000s, Australian playwrights are offering new inroads into the contemporary climate-affected world. The plays, though very different in style, form and representation, describe very specific qualities explicit to this the Great Southern Land. Yet the Gothic paradox within the emerging body of work also suggests that the climate-affected spaces and places imagined are not environmentally or ecologically unique to this country. Significantly, the plays acknowledge that climate change mitigation is an urgent priority for all nations across the world. If we don't unite to address climate change, we might all become fringe dwellers, universally belonging to and yet alienated from the planet, from the land(s) we call our own and the landscape(s) in which we live, as the places and spaces we know and love become untenable, unrecognizable and uninhabitable.

The Climate Gothic discussion posed in the preceding chapters aims to foster a climate-informed critique of the contemporary realm by exploring the current conflicting relationship between our recent experiences of climate change and our collective memory of landscape. The explorations juxtapose dramatic and theatrical perspectives that situate the sensed form and the aesthetic form within the cultural critique expressed. The development of a Climate Gothic sensibility in Australian drama is particularly relevant to contemporary socio-cultural analysis because the plays provide a cultural vocabulary and a framework for investigating human and non-human experiences in deteriorating climate-scapes. In doing so, the Climate Gothic positions climate change as the existing 'evil' in the language of cultural and social representation. Climate Gothic analysis is therefore significant to the development of Australian drama as it suggests environmental and ecological decisions and actions made in the past survive in the present, and those made in the present will carry and echo—for better or for worse—into the future of Australian culture.

Australian Climate Gothic playwrights are responding to the Anthropocene now, in the immediacy of the theatrical event or narrative, by adapting and adopting existing generic conventions of both traditional Gothic and cli-fi. The hybrid approaches aim to alert audiences to the terror, to the menacing monster that is global warming. As a result, the dramatic form is now taking responsibility to inform, discuss, argue, motivate and even empower everyday Australians to think and act, even to instigate attitudinal and/or behavioural change, no matter how large or small. The inherent performative nature of Climate Gothic drama suggests that the plays and productions provide audiences with the opportunity of witnessing and/or responding to change—change to the perpetuated cultural myths about the country that have informed our understanding of the Australian culture-scape. This shift is further informing our understanding and our expressions of the country and explorations of Nation that frame the popular and quintessential characteristics of the (white) Australian identity. The dramatic worlds defined by weather, and identified in geographic disaster and described in the specifics of Australian places, therefore offer alternative sites for new encounters with the Australian climate, which is ever transforming but still a dominant adversary in this unique culture.

Australians, overall, have always accepted environmental challenges and natural disasters as one of the offsets of living in this beautiful and haunting landscape. Since colonization, our national (white) narratives reinforce how we have worked hard to combat environmental hardship in an arid and unforgiving country that has battled hard to beat us down. However, global warming is telling us the fight will get more difficult as the battle is now one of the greatest challenges facing humanity, regardless of race, nation or environment. As one of the most arid nations in the world, Australians must come to terms with the reality of the battle ahead. While models for future climate change mitigation continue to be developed and debated, predictions of climate change's global, national, social and cultural impacts continue to alarm. The planet is doing all she can to warn us that we need to act now—that we have actually passed the tipping point. Her cries are evident in the rising temperatures and sea levels, in the habitats being destroyed at alarming rates, in the escalating death throes of species extinction, in the reality of food and water shortages and in the unpredictable and intensifying weather events savaging the planet at a rate never before seen or experienced. As a playwright, I attempt to ask audiences to heed these warnings before it is too late:

From two hundred years into the future, the SHAMAN admonishes those of us living in the Anthropocene for not listening to the Earth as she cried out in her distress.

> SHAMAN: The dust tells of what you had and of what you did. Did you not know? Tells how you kept livin' ... how some of you kept livin' in them the Days of the Final Disaster. It tells of what you didn't leave to us. Yes, yes, we knows you are full of sorrow when you cry's and cry's and cry's. It is written in the high-tongue of your ghost-speak..[1] You WEEPED in the dust. We knows. WEEPED and SORROWED and BURROWED. Too bad so sad hey?
>
> CHORUS: (*talking to the audience*) Stop. Stop that naysaying now. No good arguing. No good swearing to us that you is here alive and well. You is not here. No more. Nevermore. You is long gone. Gone ... wssshhhhhhh. We are Ghost-speaking now.
>
> SHAMAN: Is that dust I can see creeping up behind you? No? Youse seem like youse would know if youse were living in the end of days. If you'se was 'bout to be drowned in dust. Livin' in dust. You'se got the knowledge. Yes? You have. You have the knowledge to keep going? Yes? Stop the dust? Back then in the times when you could still Land-speak?[2] You had the knowledge. Yes you did. Shame you didn't do somethin' with that knowledge. Yes? (Hassall, 2015, *Dust*, Metro Arts)

I know that in writing *Dust* and indeed *Salvation* (discussed earlier in Chapters 2 and 3), my personal relationship with my beloved Australian landscape changed. Her cries got louder and my anxiety about her well-being became real. However, I have not lost hope for a positive future. The writing and productions of all the plays included in this book offer hope (even if only in some small way) in their promotion of a cause for action, change or consideration. Hope seems to be a vague and abstract concept when dealing with the problems of the Anthropocene, when we are all exhausted, anxious and despairing about the legacies we will leave future generations. However, hope *is* evident in all of the plays discussed in this book. There is hope for behavioural change in *Dust*. There is hope for the development of healthier land management strategies in *Salvation*. There is hope for improved conservation of habitat and non-human species care in *They Saw a Thylacine* and *The Dead Devils of Cockle Creek*. There is hope for a more racially united collaboration of land rights and care issues in *The White Earth*. There is hope for better support for isolated family communities in *The Flood*. There is hope for better political management of environmental issues in *The Turquoise Elephant* and finally, there is hope for the conservation of precious Australian resources in *Water Wars*.

In a cultural climate where our desert spaces, our unique places, are devalued, and homes and habitats are consistently degraded or obliterated, the plays delve deep into contemporary questions about what it is to be human and what it is to be human right now. Most significantly, they ask us to reassess our humanity, to consider our treatment of that which is 'other' and to, at the very least, empathize with non-human nature and species to strive for a planet where we can continue to cohabitate for as long as possible. While the global catastrophe is in progress and we are living in the Anthropocene's continually warming world, the Climate Gothic attempts to make sense of—or at least explore ways to—communicate our relationships with non-human nature and species in ways that science or environmental fields do not. Climate Gothic drama uses the cultural platform of the stage to synthesize factual research with speculative imagination. It can therefore extend on scientific discourse through its drawing on environmental and ecological principles to tease out emotional rather than technical responses to climate trauma. While the plays referred to in this book do not provide any concrete answers to escalating weather events, disappearing habitats, rising sea levels or the impact of colonizing processes on the land, they do suggest Australian theatre artists are taking up the challenge in including climate change as an important topic for cultural expression. We are examining our personal relationships with each other and with our much-loved place, space and nation and in doing so, we are engaging in critical discussions about the management of our unique and deteriorating landscape and disappearing peculiar Australian species and habitats:

157 days after the Day. BIRDMAN tells a story of the dust-time not too long after the Final Disaster when he received one last kiss from a beautiful woman.

CHORUS: Birds, they Disappeared.
Quakes they ripped and teared
Bitumen broke open so big that the bowels of Hell were there for all to see.

SHAMAN: Way back in them beginning times, in your time
When them hotter 'an' Hell fires scorched – your ghost-minds cracked.
It is written in Dust.

BIRDMAN: I saw her. A woman. A very pretty woman. Beautiful even. What struck me was that she was wearing lipstick. Red lipstick. And shoes. Shiny black shoes with thin heels. You know the ones?

BIRDWOMAN: I wanted to look nice/

BIRDMAN: I saw her looking over the edge into the … into the … what? What would you call it? Chasm? Emptiness? Hole? What do you call that sort of space? That space that's left behind when the earth starts to tear itself open? What do you call that?

BIRDWOMAN: Hole. End. You call it the end/

BIRDMAN: Never mind. You know what I mean. Anyway, I saw this woman looking into that space. Her dress was a mess. But her shoes were clean.
Shiny. And her lipstick was very red.

BIRDWOMAN: I found a tube in the dust. Maybelline. I could just read that.
Maybelline. I scraped the last of it out and spread it on my lips. This lipstick was the first color I've seen in a long time. It hurt my eyes/

BIRDMAN: Her lips stood out. In the Dust. Kissable. You know what I mean? I was sort of concerned that she might fall in/

BIRDWOMAN: Disappear … I could just/

BIRDMAN: So, I yelled out to her. LADY! HEY LADY! BE CAREFUL
I didn't go over to her, I just yelled out. Well yes. Yes. I was scared that it might open up more and I'd end up falling into the emptiness. So, I yelled out to her. I yelled: BE CAREFUL. And she turned and asked/

BIRDWOMAN: Did you see that?

BIRDMAN: What?

BIRDWOMAN: The bird. The bird that fell out of the sky into the … into the … into the … into the/

BIRDMAN: And I said I hadn't seen any birds for quite some time. And she said/

BIRDWOMAN: A bird just fell out. Fell out of the sky. Right out of the sky and dropped into the into the … into the/

BIRDMAN: She said it was a shame that I hadn't seen it. She said/

BIRDWOMAN: O Shame … what a shame. Such a shame you didn't see it. I think it must be the last one. The last bird ever. The last bird we'll ever see.

BIRDMAN: She thought it was a shame because she thought it was the last one.

The last bird ever. And she asked me/

BIRDWOMAN: Isn't that sad? Don't you think that's sad?

BIRDMAN: I said I supposed that I did. Before she asked me that I'd never really thought too much about birds, you know? Then she did it. She blew me a kiss. A great big red lip-sticked kiss. I tell you, I felt that kiss cut through the dust and settle on my lips. Soft like. It had been a long time since anyone had kissed me like that.

(BIRDWOMAN blows the kiss, does a graceful turn and dives into the … into the … into the … hole as described by BIRDMAN).

I was still feeling her kiss when she turned back to the hole and dived… graceful as all get out. Like a bird she was. She spread her arms wide like a bird on the wing and she dived. Into the hole. So graceful. Beautiful. The tips of those shiny black shoes pointing to the sky were the last I saw of her. I remember how pretty she was. How elegant. And the kiss. I remember her kiss ….and I regret not seeing it… the last bird… (Hassall, 2015, *Dust*, Metro Arts Production)

Notes

1. Ghost speak is a term for the stories of those facing the final encounter with climate change.
2. Land Speak is a term that means to look after and care for the earth.

References

Fuchs, E. (2002). Reading for landscape: The case for American drama. In E. Fuchs & U. Chaudhuri (Eds.), *Land/scape/theatre* (pp. 30–52). University of Michigan Press.

Hassall, L. (2015). *Dust* produced by Griffith University at the Sue Benner Theatre. Directed by Linda Hassall September 2015. Metro Arts.

Index

L. Hassall, *Theatres of Dust*,
https://doi.org/10.1007/978-981-16-6159-4

The manufacturer's authorised representative in the EU is Springer Nature Customer Service Centre GmbH, Europaplatz 3, 69115 Heidelberg, Germany. If you have any concerns regarding our products, please contact ProductSafety@springernature.com

Printed and bound by CPI Group (UK) Ltd, Croydon, CR0 4YY
30/09/2026
02225260-0009